D1258514

Integrating Productivity and Quality Management

INDUSTRIAL ENGINEERING

A Series of Reference Books and Textbooks

Editor

WILBUR MEIER, JR.

Dean, College of Engineering
The Pennsylvania State University
University Park, Pennsylvania

Additional Volumes in Preparation

Integrating Productivity and Quality Management

Johnson Aimie Edosomwan
IBM Corporation
Data Systems Division
Poughkeepsie, New York

MARCEL DEKKER, INC.   New York and Basel

Library of Congress Cataloging in Publication Data

Edosomwan, Johnson Aimie, [date]
    Integrating productivity and quality management.

    (Industrial engineering ; v. 14)
    Bibliography: p.
    Includes index.
    1. Industrial productivity--Management. 2. Quality
assurance--Management. 3. Quality of products--Manage-
ment. I. Title. II. Series.
HD56.E28 1987   658.5   86-32780
ISBN 0-8247-7657-7

MARCEL DEKKER, INC.
270 Madison Avenue, New York, New York 10016

Current printing (last digit):

10 9 8 7 6 5 4 3 2 1

PRINTED IN THE UNITED STATES OF AMERICA

Foreword

Several years ago, Johnson A. Edosomwan took my graduate course on productivity measurement, analysis, and improvement at the George Washington University. Subsequently, he completed and successfully defended his scholarly doctoral dissertation, "A Methodology for Assessing the Impact of Computer Technology on Productivity, Production Quality, Job Satisfaction and Psychological Stress in a Specific Assembly Task." In the present work he demonstrates even more fully his breadth of understanding and encyclopedic knowledge of productivity and quality issues.

In this volume, Dr. Edosomwan first clarifies basic concepts and the relationships between productivity and quality at the task and organizational levels. He then discusses the management challenges posed by rapid technological change. It is clear that rates of return, and even the continuing existence of many firms, will depend on how well they meet these challenges.

The author identifies the various levels of productivity and quality management, and presents case studies that point up various practical problems and ways of solving them. He proposes a variety of new techniques for measuring, planning, and improving productivity and quality, and sets forth step-by-step methods for implementing the techniques. Forms and checklists are provided that will be useful for practitioners.

Of special interest are Dr. Edosomwan's descriptions of tested ap-

iii

proaches for setting up formal productivity and quality management programs. He is quite aware that there are many factors besides quality that contribute to productivity improvement and thus to cost reduction. Indeed, he devotes an entire chapter to job simplification, motivation, and morale improvement in high technology environments.

This is not a work that can be read and digested in one sitting, or even in several. But it will repay students and practitioners alike to study it carefully, and keep it on their shelves for reference.

One of the rewards of teaching is having students like Johnson A. Edosomwan who go on to teach others and expand the circle of understanding in their fields. Few fields are more important than productivity improvement, which can eventually eliminate poverty from the face of the earth. Edosomwan can be proud of his contributions to the productivity and quality management literature, as are we who have worked with him.

John W. Kendrick
Scholar, American Enterprise Institute and
Professor of Economics
The George Washington University
Washington, D.C.

Foreword

This book on productivity and quality management, by Dr. Johnson A. Edosomwan, contains a valuable source of material which, when studied and applied to American industry, will have a major impact in maintaining a competitive position in the world market. Dr. Edosomwan's approach of combining quality and productivity as an integrated management team effort can only have a positive effect in the production of a quality product (or service) at minimum cost. Only by interrelating these two management functions into one all encompassing productive unit can an organization realize the most effective utilization of resources.

The conceptual framework, methodologies, and techniques for managing productivity and quality in companies and organizations is presented in a logical, step-by-step approach, which provides a means of developing and implementing a productivity and quality management program. Real-life examples resulting in significant cost reductions illustrate the sensitivity of his approach to the real world.

The information and material contained in this book provide a valuable source of information and practical applications for managers, supervisors, and employees of companies and organizations who seek to produce quality products and/or services at a minimal cost. This book is also well suited for use by students in a course work on productivity and quality management.

Carl M. Kromp, P.E.,
Professor Emeritus
University of Miami
Miami, Florida

Preface

This book goes to press at a time many companies, organizations, and government agencies worldwide are faced with the perplexing problems of inflation, international trade competition, and unemployment. Both "profit" and "nonprofit" organizations have a strong need to reduce expenses and generate acceptable levels of revenue in order to stay fiscally sound. The only way any economic entity can overcome its problems is through formal productivity and quality management.

Unfortunately, the few books that are available today have treated productivity and quality as unrelated subjects. These books have provided methodologies and techniques for addressing productivity and quality issues in one dimension. Some books emphasize the quality dimension and others emphasize the productivity dimension. Productivity and quality are interrelated and connected. The two are inseparable.

The primary objective of this book is to close the gap in the treatment of productivity and quality issues. This book provides a validated conceptual framework, methodologies, and techniques for managing productivity and quality in companies and organizations. Detailed treatment of productivity and quality methodologies, principles, and techniques is provided in the three phases of the productivity and quality management triangle (PQMT). The components of PQMT are measurement, control, and evaluation; planning and analysis; and improvement and monitoring.

This book evolved out of research, teaching, consulting, and industrial work experience in the area of productivity and quality management, and out of seminars and workshops conducted nationally and internationally. The case studies presented are from real-life experiences in both service and manufacturing organizations.

The book is intended to serve the practical needs of practitioners, researchers, consultants, students, productivity managers, quality improvement facilitators, and executives concerned with increasing productivity and quality improvement. Written in a simple manner, the book can be used as both a class text and a handbook for reference when addressing productivity and quality issues in companies and organizations.

The book is organized into eleven chapters. Chapter 1 provides definitions and concepts about the importance of productivity and quality management. The productivity and quality management triangle (PQMT) is presented. Various types of quality costs are discussed. The productivity and quality management hierarchy (PQMH) is also presented. Chapter 2 presents productivity and quality management challenges. The rapidly changing technology life cycle is also discussed. The components of the comprehensive productivity and quality management model (CPQMM) are presented. Chapter 3 discusses the connection between productivity and quality in the modern business environment. Case studies are presented to illustrate the point that productivity and quality are interrelated. Techniques are provided for understanding the basic cost of quality and cost of inspection per sample. Chapter 4 presents specific measurement models and techniques for productivity. Both the task-oriented and the technology-oriented productivity measurement models are discussed. A step-by-step implementation methodology for the models and measurement techniques is also presented. Case studies that discuss both the application of techniques and potential problems are presented. Chapter 5 discusses productivity and quality improvement through the application of statistical process control concepts. Statistical tools needed for data analysis are presented. A case study that demonstrates the application of statistical process control concepts in a group technology production environment is presented. Chapter 6 offers a conceptual framework for productivity and quality planning. Specific planning techniques are discussed under the four phases of the comprehensive productivity planning cycle. A strategy for quality planning is offered. The role of technological forecasting in the planning process is presented. Forecasting techniques are presented as tools in the overall planning process.

Chapter 7 presents the quality error removal technique (QER), which provides guidelines for work groups in resolving problems at the source of production or service. Two case studies that demonstrate the application of the QER technique are presented. Chapter 8 discusses productivity and quality improvement techniques with an emphasis on the production and service improvement technique (PASIT). A methodology for implementing the just-in-time PASIT concept is presented. Chapter 9 presents techniques for job simplification, motivation, and morale improvement. The role of technology in productivity and quality improvement is examined. Design rules that will guide systems designers in developing better systems are provided. Chapter 10 provides guidelines for promoting productivity and quality improvement programs. The common problems encountered in setting up a productivity and quality program are discussed as well as ways to counter them. A framework is provided for organizing productivity and quality management programs. A self-assessment checklist for managing productivity and quality in organizations is also provided. Chapter 11 presents conclusions on the material covered in the book. Emphasis is placed on areas of potential benefit from productivity and quality improvement. Productivity and quality management is finally explained as an on-going process that continuously seeks better techniques and methods to eliminate waste. It is believed that organizational waste can be eliminated through the application of common sense. The appendixes provide statistical tables, blank work forms for implementing productivity and quality improvement techniques, a list of productivity and quality of working life centers around the world, a glossary of terms, and a bibliography on productivity and quality management.

Johnson Aimie Edosomwan

Acknowledgments

My own ideas, knowledge, and observations on productivity and quality managment have been shaped by others who are active in this field as writers, teachers, and practitioners. I am grateful for the privilege to have been a student of or apprentice to a number of great men, such as John W. Kendrick, who can lay claim to being the father of U.S. productivity statistics and W. Edwards Deming, whose work revolutionized quality and productivity in Japan. Equally valued was apprenticeship under other esteemed professors and researchers, such as David J. Sumanth, Seymour Melman, Carl M. Kromp, Robert C. Waters, Tarek M. Khalil, Robert A. Karasek, and James N. Mosel.

My thanks to the Social Science Research Council (SSRC) of the U.S. Department of Labor and International Business Machine Corporation (IBM) for their grants, numbers 22-36-83-21 and IBM-2J2-722271-83/85, respectively. SSRC and IBM grants helped in conducting research work in productivity measurement and in gathering data for some of the case studies. My gratitude to all those companies and organizations that provided their facilities as a testing ground for the methodologies prescribed in this book. For proprietary reasons the identities of companies and organizations are masked. I am also grateful for the excellent suggestions received from the reviewers. My sincere thanks to all the Marcel Dekker staff who worked on this book with great dedication.

Many other people contributed to the planning and typing of this book. My thanks to Patricia Lopez, Gerald Talen, Edmund Hong, Kathleen Fair, Jane Barnhart, and Joan Blum, who helped on many occasions. My appreciation is expressed to my mother, Otuomagie Alice Edosomwan, and the rest of my family, who supported my early education. My love and appreciation to my wife, Mary, my daughter, Esosa, and my son, Johnson Aimie Edosomwan, Jr., for their support and encouragement. Finally, I am grateful to God for his guidance during the preparation of this book and for answering my prayers during moments of frustration.

Contents

Contents

Integrating Productivity
and Quality Management

1

Basic Concepts in Productivity and Quality Management

1.1 PRODUCTIVITY AND QUALITY IN THE MODERN BUSINESS ENVIRONMENT

In this era of technological explosion, a company or organization, regardless of its size, faces four major problems. For any project there is a limited supply of resources, such as capital, materials, energy, and labor. Further, competitive environments demand a better quality product or service at the existing price or at a lower price. Survival through acceptable profit levels requires maintaining the current market share or improving it as much as possible. In order to attain the end goal of profit, a business unit typically has multiple objectives. Problems arise when allocating scarce resources to the variety of alternative purposes competing for their use. Matching objectives with resources to attain end results is not an easy task. Only those organization that manage productivity and quality as an ongoing activity will be able to deal with these problems.

At the global level, the problems of inflation, international trade competition, and unemployment are directly affected by the level of productivity growth and the quality of goods and services. The cyclic effects of a low productivity growth rate on the national economy are shown schematically in Figure 1.1. The impact of poor quality goods and services on the survival of any economic unit is presented in Figure 1.2.

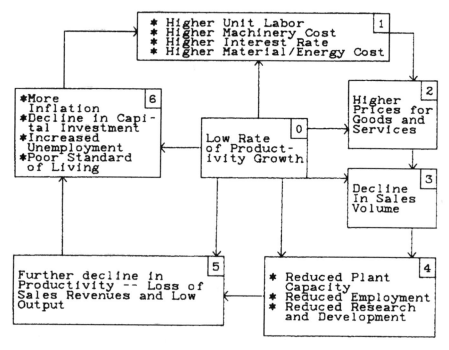

Figure 1.1   The cyclic effects of low productivity growth. (*Source*: Edosomwan, 1985b.)

In this age, every nation concerns itself with a high standard of living and better quality of work life and every organization concerns itself with high profit levels and an increased market share. For consumers who are concerned with the quality or "fitness for use" of the goods and services offered, therefore, productivity and quality management constitute the major driving force for survival.

1.2   BASIC DEFINITIONS OF PRODUCTIVITY AND QUALITY

1.2.1   Productivity

Productivity was mentioned for the first time in an article by Quesnay in 1776, and since then most authors have defined it in different ways. Table 1.1 presents a chronological classification of productivity defini-

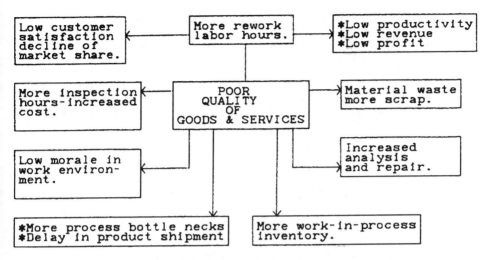

Figure 1.2   The impact of poor quality in any economic unit.

tions and measures that have been offered by both researchers and practitioners. A careful examination of these definitions reveals that some authors have defined productivity in a vague manner and others, precisely. One major similarity that could be inferred from these various definitions of productivity is that most authors viewed productivity as a "measure" of output, to one input, two inputs, or total input.

The measure also pertains to how well resources are utilized. Three different forms of productivity have been accepted universally by most researchers and practitioners. For the purpose of this text, the three forms of productivity are presented as follows (Kendrick and Creamer, 1965; Edosomwan, 1985b):

1. Total productivity is "the ratio of total output to all input factors."
2. Total factor productivity is "the ratio of total output to the sum of associated labor and capital (factor) inputs."
3. Partial productivity is the "ratio of total output to one class of input."

1.2.2   Quality

Crosby (1979) defines quality as "conformance to specification." This definition requires detailed clarification of all relevant quality charac-

Table 1.1   Definitions and Measures of Productivity in
Chronological Sequence

| Year | Author | Definition(s) | Type of productivity measure | | |
			Partial	Total factor	Total
1776	Quesnay	Mentioned productivity for the first time as relation of output to input			
1883	Littre	"Faculty to produce"			
1937	Weintraub	"Ratio of total output to the amount of labor of contractually employed workers"	X		
1955	Davis	"The change in product obtained for resources expended"			
1962	Fabricant	"A ratio of output to input, and a productivity index is always the ratio for one period (or place) relative to the corresponding ratio for another period (or place)"			
1965	Kendrick and Creamer	1. Total productivity is the "ratio of real gross output to a combination of all corresponding inputs: labor, capital, and intermediate products purchased outside the firm or industry"			
		2. Total factor productivity is the "ratio of the real products originating in the economy, industry, or firm to the sum of associated labor and capital (factor) inputs"	X	X	X
		3. Partial productivity is the "ratio of gross or net			

Year	Author	Definition(s)	Type of productivity measure		
			Partial	Total factor	Total
		output to one class of input"			
1970	Dewitt	(Personal productivity) = revenues per employee: net earnings per employee; (capital productivity) = revenues per stock holders equity \$; operating income per stock holders equity \$; (facilities productivity) = revenues per plant and equipment \$; operating income per plant and equipment \$	X		
1971	Soliman and Hartman	"Productivity is a particular type of behavior within an organization. It may be high, a positive and desirable behavior; or low, a negative and undesirable behavior"			
1972	Craig and Harris	Total productivity is a "ratio of total output to total input"			X
1976	Siegel	"Family of ratios of output to input			
1976	Hines	"Ratio of total output to total input			X
1976	Mundel	Productivity index = $\dfrac{\text{outputs}}{\text{Inputs}} \times 100$			X
1977	Ross	"The ratio of some measure of output to some measure of input—a difficult measure at best,			

(Continued)

Table 1.1 (*Continued*)

Year	Author	Definition(s)	Partial	Total factor	Total
		is usually attributed to the improved efficiency of some specific resource such as capital, money, materials, or technology"			
1977	Taylor Davis	Total factor productivity = $\dfrac{\text{total value} - \text{added output}}{\text{total input (capital and labor)}}$		X	
1978	Mali	"Productivity is the measure of how well resources are brought together in organizations and utilized for accomplishing a set of results. Productivity is reaching the highest level of performance with the least expenditure of resources."		X	X
1978	Stewart	"As the ratio of performance towards organizational objectives to the totality of input parameters"			X
1979	Sumanth	"Ratio of total tangible output (in physical or value terms) to the sum of all the tangible inputs (in cost terms)"			X
1985	Edosomwan	Total productivity is a "ratio of all measurable output (finished units, partial units, and other outputs associated with quantity produced) to the sum of all measurable inputs			X

| | | | Type of productivity measure | | |
| | | | | Total | |
Year	Author	Definition(s)	Partial	factor	Total
		(computer expenses, robotics expenses, labor, materials, energy capital, data processing expenses, and other administrative expenses)"			

teristics and total evaluation and understanding of the entity involved. Quality when viewed as conformance to specification has great potential for being a very effective business strategy. However, this definition can result in a lack of commitment to quality by members of an organization that does not acknowledge the prescribed specification. There is also a danger of adherence to the specification even if it is not the most productive method to accomplish the desired end.

Juran (1979) defines quality as "fitness for use." The unique aspect of this definition is the inclusion of the concept of the user. Quality is viewed as one that requires every member of the organization to provide the next person in the process with an acceptable product or service. This means that everyone is responsible to perform a task in such a manner that the product can be used immediately and in the most efficient manner possible. Juran further identified four important parameters of fitness for use as

1. Quality of design
2. Quality of conformance
3. The abilities
4. Field service

The interrelationships among these parameters are presented in Figure 1.3.

Quality is a term that has valuable meaning to both producer and customer, as shown in Figure 1.4. The producer views fitness for use in terms of the ability to process and produce with less rework, less scrap, minimal downtime, and high productivity. From the customer's view-

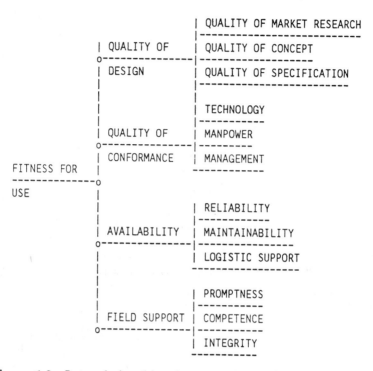

Figure 1.3   Interrelationships between fitness for use parameters. (*Source:* Juran, 1979, pp. 2-9.)

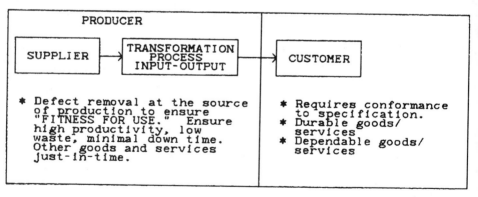

Figure 1.4   Supplier, producer, and customer fitness for use relationships.

point, fitness for use includes product durability and availability of spare parts.

There are two other aspects of quality: the quality of design and the quality of conformance. Quality of design pertains to quality obtained through changes or manipulation in design parameters. The differences in quality are a result of differences in size, materials used, tolerances in manufacturing, reliability, equipment utilized, and temperature. Quality of conformance is a measure of how well the product conforms to the specifications and tolerances required by the design. Such factors as training, the production process, motivation levels, procedures, and quality assurance systems can have an effect on the quality of conformance.

Quality Characteristics

These are elements of fitness for use that typify the variety of uses of a given product. Quality characteristics may be of several types:

1. Time-oriented (serviceability, reliability, maintainability)
2. Sensory (color, taste, beauty, appearance, and others)
3. Structural (including frequency, weight, length, and viscosity)
4. Commercial (warranty)
5. Behavioral or ethical (fairness, honesty, courtesy, and so on)

Improvements in quality are therefore made by examining the design and conformance phases with their associated characteristics through the application of techniques and methods (changes in design, timing, inspection procedures, process control procedures, and so on).

1.3   FORMAL DEFINITION OF PRODUCTIVITY AND QUALITY MANAGEMENT

Productivity and quality management will be defined as follows. Productivity and quality management is an integrated process involving both management and employees with the ultimate goal of managing the design, development, production, transfer, and use of the various types of products or services in both the work environment and the marketplace. The process requires the total involvement of everyone in the planning, measurement, evaluation, control, and improvement of productivity and quality at the source of production or service center.

1.4   THE PRODUCTIVITY AND QUALITY
MANAGEMENT TRIANGLE

The productivity and quality management triangle (PQMT) shown in
Figure 1.5 encompasses an information system that provides input of
information relevant to the planning, performance, and measurement
processes and implementation of corrective actions and techniques
that improve productivity. For an organization initiating comprehensive productivity and quality management, measurement, control, and
evaluation is the first stage in the PQMT triangle. Once a set of
measures has been developed, plan targets and strategies are formulated. Based on these strategies, short-range objectives are formulated and operational improvement techniques are then used to implement short- and long-range objectives. In order to assess the extent
to which the improvements were successfully implemented, measurements, control, and evaluation are performed again. This triangular
relationship thus continues as long as the comprehensive productivity
and quality program exists in the organization.

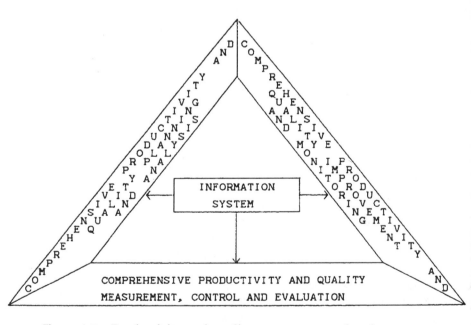

Figure 1.5   Productivity and quality management triangle .

1.5 BENEFITS OF PRODUCTIVITY AND QUALITY MANAGEMENT

The management of productivity and quality can provide the following benefits to organizations.

1. Productivity and quality management will enable the consumer to pay low prices for goods and services because the cost of production is reduced owing to reduced rework and gains in productivity.
2. Productivity and quality management enable the effective utilization of resources. More goods and services are produced for a reasonable amount of expended resources.
3. Productivity and quality management provide the basis for higher real earnings for employees. The reduction in the cost of production of goods and services can allow increases in wages without significantly offsetting gains in total productivity.
4. Productivity and quality management components (planning and analysis, measurement, evaluation and control, and improvement and monitoring) provide an organization with the strength to deal with internal operating weakness and external competition.
5. Productivity and quality management will enable an organization to be more profitable because quality improvement results in reduced rework, reduction in scrap, better utilization of tools and equipment, and less work in process inventory, which in turn leads to higher productivity. (Profits = revenue from sales − cost of manufacture.) Minimization of the cost of manufacture for goods and services improves the profit margin. The production of good quality products also improves the market share because the satisfied customer will buy more and will recommend the product to another consumer.
6. Productivity and quality management enable the public to realize greater social benefits through increased public revenues from organizations.

1.6 TOP MANAGEMENT ROLE IN PRODUCTIVITY AND QUALITY MANAGEMENT

Deming (1982, pp. 13–59) discusses what top management must do to improve both productivity and quality. The fourteen points offered by

Deming have been successfully implemented in Japan and most of the industrialized countries in the world. A summary of Deming's fourteen points follows.

1. Create constancy of purpose toward improvement of product and service, with a plan to become competitive and to stay in business. Decide to whom top management is responsible.
2. Adopt the new philosophy. We are in a new economic age. We can no longer live with commonly accepted levels of delays, mistakes, defective materials, and defective workmanship.
3. Cease dependence on mass inspection. Require, instead, statistical evidence that quality is built in to eliminate the need for inspection on a mass basis. Purchasing managers have a new job and must learn it.
4. End the practice of awarding business on the basis of price tag. Instead, depend on meaningful measures of quality, along with price. Eliminate suppliers that cannot qualify with statistical evidence of quality.
5. Find problems. It is management's job to work continually on the system (design, incoming materials, composition of material, maintenance, improvement of machines, training, supervision, and retraining).
6. Institute modern methods of training on the job.
7. Institute modern methods of supervision of production workers. The responsibility of foremen must be changed from sheer numbers to quality. Improvement of quality will automatically improve productivity. Management must prepare to take immediate action on reports from foremen concerning such barriers as inherited defects, machines not maintained, poor tools, and fuzzy operational definitions.
8. Drive out fear so that everyone may work effectively for the company.
9. Break down barriers between departments. People in research, design, sales, and production must work as a team, to foresee problems of production that may be encountered with various materials and specifications.
10. Eliminate numerical goals, posters, and slogans for the work force that ask for new levels of productivity without providing methods.
11. Eliminate work standards that prescribe numerical quotas.

12. Remove barriers that stand between the hourly worker and the right to pride of workmanship.
13. Institute a vigorous program of education and retraining.
14. Create a structure in top management that will push every day on the other thirteen points.

1.7 CAUSES OF PRODUCTIVITY DECLINE IN COMPANIES

Mali (1978, pp. 24–32) presented twelve causes for the decline of productivity in organizations:

1. The inability to measure, evaluate, and manage the productivity of white-collar employees can cause a shocking waste of resources.
2. Rewards and benefits given without requiring the equivalent in productivity and accountability causes spiraling inflation.
3. Diffused authority and inefficiency in complex organizations cause delays and time lags.
4. Organizational expansion that lower productivity growth results in soaring costs.
5. There is low motivation among a rising number of affluent workers with new attitudes.
6. Late deliveries may be caused by schedules that have been disrupted by scarcity of materials.
7. Unresolved human conflicts and difficulties in team work result in the firm's ineffectiveness.
8. Increased legislative intrusions or antiquated laws result in constrained management options and prerogatives.
9. Overspecialization in work processes results in monotony and boredom.
10. Rapid technological changes and high costs result in a decline in new opportunities and innovations.
11. Increasing demand for leisure time causes disruption of time commitments.
12. A practitioner is unable to keep pace with the latest information and knowledge.

Although Mali's points focus on productivity, much the same reasons are applicable to the decline of quality within organizations.

1.8   QUALITY COSTS

Quality costs are those various categories of costs that are associated with identifying, avoiding, producing, maintaining, and repairing products that do not meet specifications. Quality costs are in four categories. These and some examples are shown schematically in Figure 1.6. Each quality cost category will now be described.

1.8.1   Prevention Costs

Prevention costs are associated with designing, implementing, and maintaining a quality system capable of anticipating and preventing quality problems before they generate avoidable costs. All costs related to efforts in design and manufacturing that are directed toward the prevention of nonconformance are in this category.

1.8.2   Appraisal Costs

Appraisal costs are associated with measuring, evaluating, or auditing products, components, and purchased materials to assure conformance to quality standards and performance specifications.

1.8.3   Internal Failure Costs

Internal failure costs are associated with defective products, components, and materials that fail to meet quality requirements, and this failure is discovered prior to delivery of the product to the customer.

1.8.4   External Failure Costs

External failure costs are generated by defective products shipped to customers. The product generally does not conform to requirements and fails to meet the satisfaction of the customers.

1.9   EFFECTIVE SYSTEM MANAGEMENT
FOR PRODUCTIVITY
AND QUALITY IMPROVEMENT

In order to improve productivity and quality, an effective management system must be in place. Such a management system requires applying common sense and creativity in making the required changes in

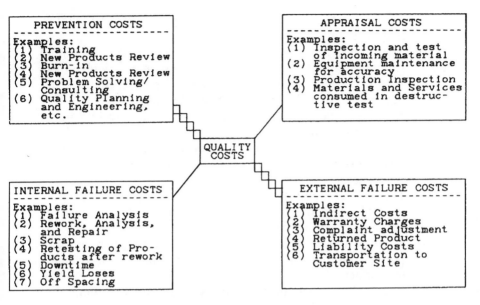

Figure 1.6   Categories of quality costs.

policies, organizational structure, and procedures that allow a fertile environment for productivity and quality improvement. The following are basic requirements.

1. Management philosophy must be one that is geared toward continuous productivity and quality improvement at the source of production or service. Management of productivity and quality is not a one-time action. It is a continuous process that involves both management and employees, in producing and delivering good products and services.
2. Management and employee commitment to quality and productivity improvement can be translated into a written policy. There must be documentation of such a policy and clear specifications of all the necessary details to assure it.
3. Management philosophy must accept no levels of defect. Low productivity is not acceptable.
4. There must be a system in place for measurement, control and evaluation, planning and analysis, and improvement and monitoring of all activities.

5. There must be a data base for the storage and retrieval of information about the process.
6. There must be a clear definition of process parameters and variables. Input and output elements must be clearly identified.
7. All process parameters must be characterized with complete understanding of repeatability and variability.
8. There must be process ownership by everyone involved in the management process.
9. Productivity and quality improvements based on the implementation of key actions must be documented and rewarded on a timely basis.
10. The management information system must provide a real time feedback mechanism to everyone involved in the productivity and quality improvement efforts.
11. There should be a provision for ongoing management and employee training in productivity and quality improvement techniques.
12. There should mechanisms for the identification and removal of barriers between functional departments to facilitate open communication.
13. Management must provide clear guidance and clear goals and objectives, specify short- and long-term commitments, and set the direction for excellence.
14. The attainment of a good quality product should be through prevention of defects, not through detection of defects.
15. The removal of defects from a process should be done through root cause analysis and the implementation of proper corrective action. Cost inspections should be discouraged. The strategy should be that which utilizes sampling inspection when necessary and no inspection when it is not warranted.

1.10 PRODUCTIVITY AND QUALITY MANAGEMENT HIERARCHY

In order for a productivity and quality management system to provide useful and meaningful results, it is important for the level of emphasis to be clearly specified and understood within the productivity and quality management hierarchy (PQMH) shown in Figure 1.7. The productivity and quality management hierarchy suggests that productivity

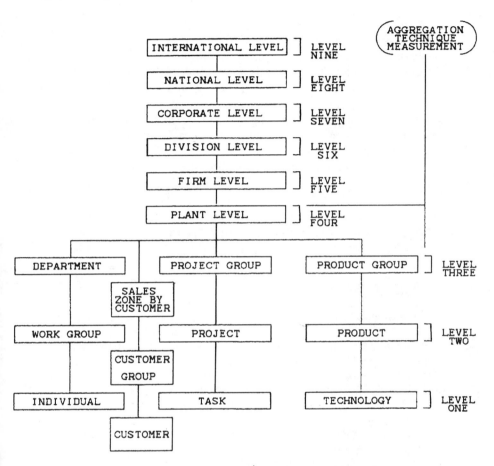

Figure 1.7 Productivity and quality management hierarchy.

and quality management should start at the basic levels (individuals, task, and technology). If all the associated inputs and outputs are clearly identified and managed properly at this level, the results have tremendous impact. It becomes less cumbersome to quantify and manage productivity and quality at the other eight different levels. The PQMH also provides a basis to identify the productivity and quality issues from a global perspective. Most importantly, it shows that the productivity and quality management of any nation starts from the basic units. At each level of the PQMH, techniques and methodology are required for the planning and analysis process, performance and measurement process, and improvement and monitoring process. Aggregation technique is usually applied to understand productivity and quality management issues at the higher levels (levels 4–9). It is usually helpful if there are consistent and conducive policies affecting productivity and quality at the higher levels of the PQMH. In Japan, for example, there is partnership between the government and industry on issues that affect productivity and quality.

In the next chapter, the components of productivity and quality management will be presented after a discussion of productivity and quality management challenges. It will also be show that the one major issue that poses the greatest challenge is our ability to effectively manage productivity and quality during dynamic change in the technological life cycle.

REFERENCES

Business Week, 1982. Quality: The U.S. drives to catch up, November 1.

Craig, C. E., and C. R. Harris, 1972. Productivity concepts and measurement—a management viewpoint. Master's thesis, MIT, Cambridge, Massachusetts.

Craig, C. E., and C. R. Harris, 1973. Total productivity measurement at the firm level. *Sloan Management Review*, Vol. 14, No. 3, pp. 13–39.

Crosby, P. B., 1979. *Quality is Free.* Signet, New York.

Davis, H. S., 1955. *Productivity Accounting.* University of Pennsylvania Press, Philadelphia.

Deming, E. W., 1982. *Quality, Productivity, and Competitive Position.* MIT Press, Cambridge, Massachusetts.

Dewitt, F., 1970. Technique for measuring management productivity. *Management Review*, Vol. 59, pp. 2–11.

Edosomwan, J. A., 1980. Implementation of the total productivity model in a manufacturing company. Master's thesis, Department of Industrial Engineering, University of Miami, July.

Edosomwan, J. A., 1985a. A methodology for assessing the impact of computer technology on productivity, production quality, job satisfaction and psychological stress in a specific assembly task. Doctoral dissertation, Department of Engineering Administration, The George Washington University, Washington, D.C. 20052, January.

Edosomwan, J. A., 1985b. A task-oriented total productivity measurement model for electronic printed circuit board assembly. International Electronic Assembly Conference Proceeding, October 7–9, Santa Clara, California.

Edosomwan, J. A., 1986a. A conceptual framework for productivity planning. *Industrial Engineering*, January.

Edosomwan, J. A. 1986b. Statistical process control in group technology production environment. SYNERGY '86 Proceedings, June 16–18, Universal City, California. Sponsored by Society of Manufacturing Engineers, Computer and Automated Systems Association, and the American Production and Inventory Control Society.

Edosomwan, J. A., 1987. The meaning and management of productivity and quality. *Industrial Engineering*, January.

Fabricant, S., 1962. Which productivity? Perspective on a current question. *Monthly Labor Review*, Vol. 86, No. 6, pp. 609–613.

Fabricant, S., 1969. *A Primer on Productivity*. Random House, New York.

Hines, W. W., 1976. Guidelines for implementing productivity measurement. *Industrial Engineering*, Vol. 8, No. 6.

Juran, J. N., 1979. *The Quality Control Handbook*. McGraw-Hill, New York.

Juran, J. M., and F. M. Gryna, Jr., 1980. *Quality Planning and Analysis*. McGraw-Hill, New York.

Kendrick, J. W., in collaboration with the American Productivity Center, 1900. *Improving Company Productivity. Handbook with Case Studies*. The John Hopkins University Press, Baltimore.

Kendrick, J. W., and D. Creamer, 1965. *Measuring Company Productivity: Handbook with Case Studies*. Studies in Business Economics, No. 89, National Industrial Conference Board, New York.

Mali, P., 1978. *Improving Total Productivity: MBO Strategies for Business, Government and Not-for-Profit Organizations*. John Wiley and Sons, New York.

Mundel, M. E., 1976. Measures of productivity. *Industrial Engineering*, Vol. 8, No. 5, pp. 24–26.

OEEC, 1950. *Terminology of Productivity*. OEEC, 2,2 rue Andre-Pascal, Paris 16.

Quality Magazine, 1981. *Quality, a management gambit*, June.

Siegel, I. H. *Measurement of Company Productivity in Improving Productivity Through Industry and Company Measurement*. (National Center for Productivity and Quality of Working Life, Series 2), U.S. Government Printing Office, Washington, D.C., 1976.

Siegel, I. H., 1980. *Company Productivity: Measurement for Improvement*. The W. E. Upjohn Institute for Employment Research, Kalamazoo, Michigan, April.

Stewart, W. T., 1978. A yardstick for measuring productivity. *Industrial Engineering*, Vol. 10, No. 2, pp. 34–37.

Sumanth, D. J., 1979. Productivity measurement and evaluation models for manufacturing companies. Doctoral dissertation, Illinois Institute of Technology, Chicago, August. (University Microfilms, Ann Arbor, Michigan, No. 80-03, 665).

Taylor, B. W., III, and R. K. Davis, 1977. Corporate productivity—getting it all together. *Industrial Engineering*, Vol. 9, No. 3, pp. 32–36.

2

Productivity and Quality
Management Challenges

This chapter discusses manifestations of the productivity and quality problem in the modern business environment. The key productivity and quality management challenges are discussed, and there is a presentation of the components of the comprehensive productivity and quality management model.

2.1 THE PRODUCTIVITY AND QUALITY PROBLEM

Although several factors contribute to the economic decline and the consequent standard of living, there seems to be general agreement that the productivity level and growth rate and the quality of goods and services can have a major impact on the national economy. For example, during the first half of this century, the United States maintained productivity leadership in the world with a fourfold labor productivity increase. However, since World War II, productivity growth in the United States has declined considerably.

The data complied by Kendrick (1965) and Bowen (1979) reveal an awesome decline in the annual productivity growth rate in the U.S. private business sector: from 3.2% in 1947–1966, to 2.1% in 1966–1973 and 0.8% in 1973–1979. As shown in Figure 2.1, Rahn also presented a similar trend of labor productivity growth in the United States business

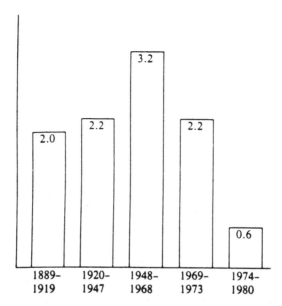

Figure 2.1 Average annual labor productivity growth in the United States private business sector, 1889–1980. (*Source*: Rahn et al., 1981)

sector from 1889–1980. The most recent projection by Kendrick (1984) shows a moderate increase of about 2.5% between 1984 and 1986.

Between January 1985 and August 1986, several major corporation cut back the number of workers to reduce their losses and to stay in business. Table 2.1 presents cutback statistics from some of the major corporations. Laying off workers is not the answer to productivity and quality problems. The answer lies in an ongoing improvement process that ultimately requires integration of all the elements of the work organization to achieve the elimination of waste, reduction of cost, and defect-free product through implementation of better techniques and working conditions. Although the productivity and quality problem in the United States may be unique to its economic climate, a careful look at other countries will reveal that similar problems and trends exist in other countries as well. Kendrick (1984) presented a productivity comparison between the United States and the rest of the world. Although

Table 2.1 Cutbacks in Major Corporations Between January 1985 and August 1986

Company	Number of Employees	% Work Force
Apple	1,200	20
ARCO	6,000	18
AT&T	35,251	10
CBS	1,175	14
Cessna	1,000	21
Combustion Engineering	7,300	20
Dupont	12,000	11
Kodak	13,700	10
Exxon	6,900	17
General Electric	26,000	8
Greyhound	3,000	21
Intel	2,600	10
Polaroid	750	5
Union Carbide	8,000	8
Wang	3,200	10

Source: Business Week, 1986, p. 45.

the quality of goods and services has received unprecedented attention in recent years, expert opinions indicate there is a lot to be done in improving quality in all industrial sectors. Juran (1981) presented his perception of the relative quality of goods and services produced by the West and those produced by Japan. As shown in Figure 2.2, Japan is shown to have a leading edge in quality. In 1980, the American Society for Quality Control conducted a survey using 10,000 questionnaires to determine the opinions of the heads of household on the quality of the goods and services produced in the United States. The response rate to the questionnaire was 71%. The results of this survey are shown in Figure 2.3. The results of this survey indicate that quality deserves significant attention in all industrial sectors. Holusha (1984) stated that the quality issue has major implications for the health of the domestic (automobile) industry, which reported a combined loss of more than $4 billion in 1980 and laid off more than 200,000 workers.

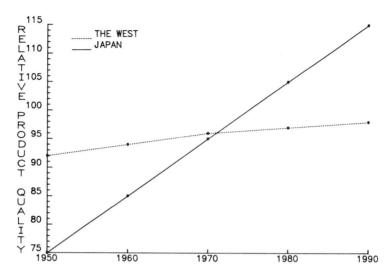

Figure 2.2   Relative product quality comparison between Japan and the West. (*Source*: Juran, 1981, p. 16)

2.2   PRODUCTIVITY AND QUALITY MANAGEMENT CHALLENGES

Edosomwan (1986e) stated that as the current explosion of technology continues, organizations will be faced with the following productivity and quality management challenges:

Managing the Dynamic Change in Technology Life Cycle to Improve Productivity and Quality

As shown in Figure 2.4, the various types of technology used to improve productivity and quality in the early 1960s had an average life cycle of ten years. From the producer, supplier, and consumer perspective, the ten-year life span for a typical technology was still short before new technologies were introduced into the work environment. By 1986, most technology life cycles have shortened to about two years and are projected to be six months by the year 2000. This continually shortening life cycle poses a real challenge for all organizations that desire to use technology as a mechanism for productivity and quality improvement. It requires that the change from one product type to another will be

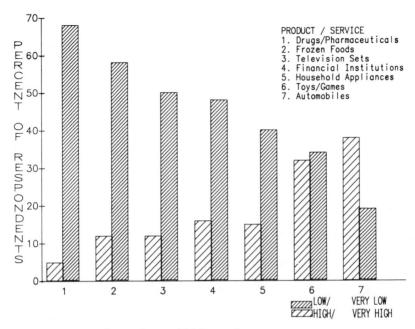

Figure 2.3   Quality ratings of U.S. products.

done at a much faster pace. Product model development, specification monitoring, prototype testing, product manufacturability and serviceability, and process qualification and improvement will have to be performed within a short time. Just-in-time productivity and quality improvement will require new techniques and tools for managing the dynamic change of the input and output elements in the work environment. It also requires a very close working relationship between the supplier, producer, and consumer of the goods and services.

Emphasis on Quality at the Source of Production and Service

For several decades, the emphasis has always been on inspecting quality into a product and service. As a result, almost every process within the work organization has an inspection station. In addition to adding to the cycle time of a product or service, inspection stations add operations that also create additional rework, which in turn increases the cost of production or service. By the year 2000, the shortened life cycle of technologies will not allow enough time for work-in-process in-

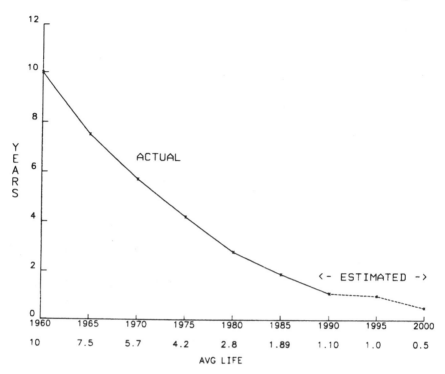

Figure 2.4   Average life cycle trend of a typical technology. (*Source:* Edosomwan, 1986e)

spections or inspections after each process. Quality will be needed at the source of production. All the mechanisms needed for a transformation process must perform correctly the first time with high reliability.

Technical Vitality Requirements for Productivity and Quality Management

The complexity of the production and service processes will increase with increasingly sophisticated automation. The new technologies will require high-level skills for design, implementation, and maintenance. The productivity and quality management will require advanced interdisciplinary knowledge, with strong demand skills that will act as a systems integrator (SI).

Motivating Workers to Increase Productivity and Improve
Quality in the Work Environment

Although several motivational techniques exist today, the effectiveness of the existing techniques may not be adequate in this era of technological explosion. New motivational techniques that address the changing values and role of workers will always be needed. More emphasis will have to be placed on task enlargement, task evaluation, psychological evaluation, morale management in real time, and other physiological needs of workers.

Assessing and Measuring the Impact of Technology
on Productivity and Quality

The rate of return on investment (ROI) and the net present value (NPV) method have been the most widely used approaches for justifying the productivity of new technologies. Defects per unit produced has also been used to justify the quality of new technologies. In the era of expanded technology, the highly capital-intensive work environment requires a comprehensive approach for justifying the productivity and quality of new technologies before and after their introduction into the work environment. Such a comprehensive approach must consider the total cost of doing business. In Chapter 4 a total productivity measurement model is presented. This model is useful for justifying new technologies.

Managing the Impact of Group Technology Application

As shown in Figures 2.5 and 2.6, the layout of machines and tools is traditionally scattered and specialized. In recent years, some organizations have realized benefits in productivity and quality improvements through group technology (GT) applications. However, the group technology layout shown in Figure 2.7 requires ongoing process control and monitoring. It is a challenge to balance the parameters and variables among processes, work cells, and machines. As GT applications receive more emphasis, appropriate productivity and quality management techniques are needed.

Planning and Analysis of Production and Service Variables

To properly manage productivity and quality in a dynamic technology-oriented work environment, the input and output components and the

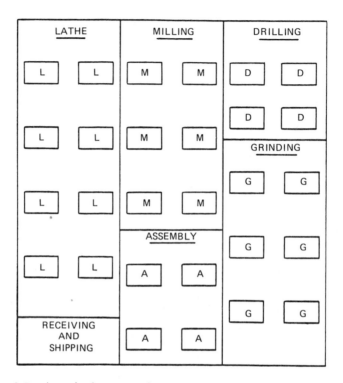

Figure 2.5   A typical process layout.

factors affecting them must be carefully planned. The total cost of doing business must be analyzed periodically to continue to minimize product or service cost overruns. Emphasis should be placed on using comprehensive productivity and quality planning techniques in both manufacturing and service organizations.

Managing the Loop Between the Supplier, Producer, and Consumer

Dynamic change in work processes and the technology life cycle brings about a shorter response time for all production and service requirements. Real time management of productivity and quality requires the elimination of excessive transportation, repetition of tasks, and delays between the supplier, producer, and consumer. The network

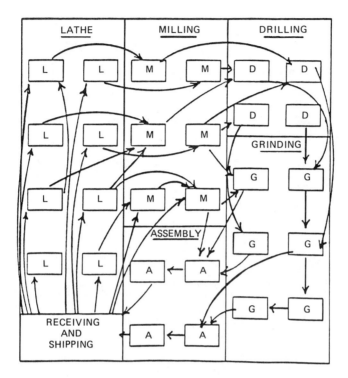

Figure 2.6   A typical process layout work flow pattern.

of information that is the supplier, producer, and consumer loop must be achieved by the year 2000. Cost minimization criteria, service turnaround time, error-free product, and service guarantee must be applied to vendor selection. For each process, material and other requirements must be delivered in time to facilitate real time productivity and quality improvement.

Ongoing Search for Productivity and Quality Management Techniques and Methodologies

The need exists for additional techniques and methodologies for improving, monitoring, planning, analyzing, measuring, and evaluating productivity and quality in a sophisticated high-technology work environment. Useful techniques for productivity and quality improvement are developed as the work organization changes.

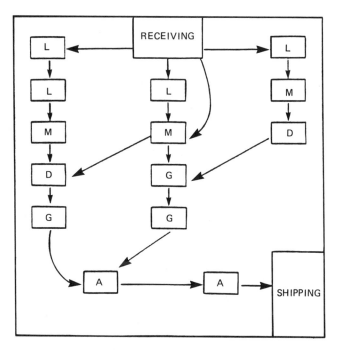

Figure 2.7   A typical group technology layout and work flow pattern.

Managing the Changing Values and Composition of the Work Force

As more attention is given to productivity and quality improvement through technological development and implementation, the values of the workers are bound to change with the new technology. How the quality of working life affects productivity and quality improvement will have to be addressed. New, flexible working conditions may have to be adopted to accommodate the changing values of men and women in a technologically advanced society. Technological advancement increasingly makes it possible for many to work at home. However, unless on-line process control linkages are developed for the sophisticated information network, system downtime is bound to have a significant negative impact on the productivity and quality of goods and services.

Effective Organization Structure for Productivity
and Quality Management

The hierarchical organization is the most widespread in the business environment. Hierarchical organizations have stability in their favor. They provide efficient mechanisms for coordinating large numbers of people through simultaneous central direction and specialization. However, the multilayers of management within the hierarchical structure can sometimes be an impediment to the quick resolution of problems within a work organization. The bureaucratic delays and bottlenecks created by the various multilayers of management are not adequate for an environment with a dynamic need for productivity and quality improvement. By the year 2000, a typical economic entity will experience dynamic change and disturbance primarily driven from the shortened technology life cycle. To quickly respond to productivity and quality management, an alternative organizational structure that utilizes fewer levels of management layers will be sought. The new form of organizational structure may be viewed as a flattened hierarchical structure with minimal levels of management. The main benefit to be derived from such a structure is a quick response rate to people, productivity, and quality issues. It will facilitate clarity and economies of scale, direct the vision of individuals and managerial units toward performance, and improve the day-to-day decision-making process.

Better Coordination of the Productivity and Quality
Management Hierarchy

For our tools and techniques to improve productivity and quality, the state or level of application within the productivity and quality management hierarchy (PQMH), specified in Chapter 1, must be carefully understood. The coordination of activities between each level of the PQMH requires the total involvement of everyone in the prescribed work environment. Figure 2.8 illustrates the total teamwork involvement in the productivity and quality management process.

2.3 COMPONENTS OF THE COMPREHENSIVE PRODUCTIVITY AND QUALITY MANAGEMENT (CPQM) MODEL

The comprehensive productivity and quality management model components shown in Figure 2.9 encompass a management information

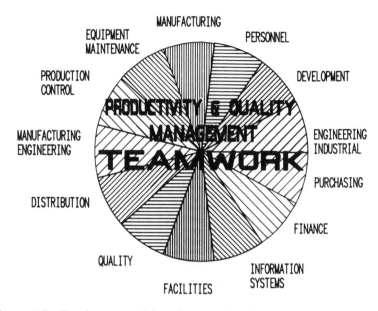

Figure 2.8   Total teamwork involvement in the productivity and quality management process.

system and a feedback mechanism that provide the necessary information about production or service process characteristics and variables, process performance through measurement, action on the process inputs, the transformation process, and action on the output. To improve productivity and quality, all production and service parameters must be characterized with complete understanding of repeatability and variability. There must be a system for real time measurement, control, evaluation, planning, and improvement. Correction of problems and defect removal should be done through system-aided root cause analysis. To avoid excessive inventory problems, the system control and balancing should be used to ensure that the rate of production or services is equivalent to the rate of consumption. However, the balancing acts should be done with the awareness of fallout rates from the input process, transformation process, and output management process.

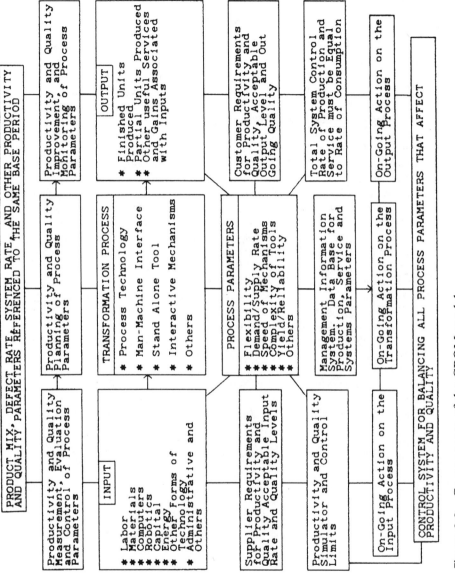

Figure 2.9   Components of the CPQM model.

REFERENCES

Bowen, W., 1979. Better prospects for our ailing productivity. *Fortune*, December 3, pp. 68–76.

Edosomwan, J. A., 1985. Computer-aided manufacturing impact on productivity, production quality, job satisfaction, and psychological stress in an assembly task. Proceedings for the Annual International Industrial Engineering Conference, December.

Edosomwan, J. A., 1986a. A conceptual framework for productivity planning. *Industrial Engineering*, pp. 64–69. January.

Edosomwan, J. A., 1986b. Managing technology in the workplace: A challenge for industrial engineers. *Industrial Engineering*, February, pp. 14–18.

Edosomwan, J. A., 1986c. The impact of computer-aided manufacturing on total productivity. Proceedings of the 8th Annual Conference on Computers and Industrial Engineering, Orlando, Florida, March.

Edosomwan, J. A., 1986d. Statistical process control in group technology production environment. The Conference on Functional Interfacing for Computer Integrated Manufacturing, SYNERGY '86 Proceedings APICS and SME, Universal City, California, June 16–18.

Edosomwan, J. A., 1986e. Productivity and Quality Management—A challenge in the year 2000. Working paper for the Annual International Industrial Engineering Conference, Boston, Mass. December.

Holusha, J., 1984. Quality Woes Bedevil Detroit, Gain Made, But Japan Is Still Seen as Leader. *The New York Times*, April 30, p. D-1.

Juran, J. N., 1981. Product quality—A presentation for the west. *Management Review*, American Management Association, June/July, p. 16.

Kendrick, J. W., 1984. In collaboration with the American Productivity Center. *Improving Company Productivity. Handbook with Case Studies*. The Johns Hopkins University Press, Baltimore and London.

Kendrick, J. W. and D. Creamer, 1965. *Measuring Company Productivity: Handbook with Case Studies*. Studies in Business Economics, No. 89. National Industrial Conference Board, New York.

Rahn, R. W., et al., 1981. *Productivity, People, and Public Policy*. U.S. Chamber of Commerce, Washington, D.C.

3

The Productivity–Quality Connection

This chapter examines the connection between productivity and quality in the modern business environment. Examples from the literature and case studies are presented to illustrate that productivity and quality are connected, interrelated, and inseparable. It is also shown that productivity increases as quality improves. Basic mathematical expressions are provided for the computation of the cost of quality.

3.1 THE CONNECTION

Midas (1981) stated that the key to actual productivity improvement is understanding the elements that can bring about productivity growth. One of these is the important leverage quality strategy can have in productivity improvement. Danforth (1984) pointed out that too many people think that high quality always cost more: this is wrong. Producing more—inefficiently—at the expense of quality is no way to increase productivity. Putting more inspectors on the line to find mistakes is the wrong approach. Doing something over again that was not done right the first time decreases efficiency, wastes money, and lowers productivity. Fiegenbaum (1979) estimated that poor quality can represent about 15–40% of the plant's productive capacity. Cole (1981) estimated the impact of implementing the "find and fix" philosophy in the automobile industry by stating: "An estimated one out of every ten workers in automobile assembly plants in the United States is engaged in repairing substandard items, resulting in truly staggering costs for scrap, rework, retest, downtime, yield losses, and disposal of substandard components." Cole further explained that another inefficiency that results from this philosophy is that it increases the number of inspectors required to detect defects. The author defined inspection as an operation that adds to the cost of a product. He estimated a typical ratio of one inspector to twenty operators in U.S. factories and one inspector to thirty operators in Japan. Cole pointed out that inspection detracts from productivity and that reliance on inspectors causes employees to be less concerned about quality, which in turn leads management to hire even more inspectors. Indeed, problems are passed on to the dealer, who receives larger dealer preparation fees to make adjustments and corrections. The consumer often receives a costlier product of poor quality. This cycle reveals the disadvantages of separating quality assurance from its execution.

3.2 CASE STUDY: A PRINTED CIRCUIT BOARD MANUFACTURING ENVIRONMENT

3.2.1 Case Study Description

A manufacturing company that produces printed circuit boards (PCB) to customer order was interested in improving the quality of the PCB. The parts and components used for the assembly of the boards were from different suppliers. The parts and components were assembled in a group machining process at a certain rate of PCB per hour in each machine group cell. The PCB production process operated at a yield of 63%. The proportion of the output that does not nonconform is 37%. Owing to high demand for reliable PCB, the firm can no longer afford to ship defective PCB to the customer. To resolve the problem, the firm instituted eight inspection and rework processes. The direct and indirect manufacturing cost due to inspection and rework was approximately $288,000 per week. The increased inspection and rework stations created additional problems: increased product cycle time, more work-in-process inventory, decline in labor productivity, increased handling, and additional defects created by the inspection in rework process itself. These problems also affected the overall effective utilization of such resources as labor and energy. The profit margin of the firm was also affected significantly.

3.2.2 Improvement Methodology and Results

A flow diagram of the PCB manufacturing process is shown in Figure 3.1. Careful analysis revealed that there was no control in place to monitor the production process. Control charts were developed and implemented in all operations. The statistical process control mechanisms enabled the process performance to be monitored properly. Corrective actions were taken to ensure good quality at the source of production. Data gathering and analysis were collected in real time. The implementation of statistical process control led to significant im-

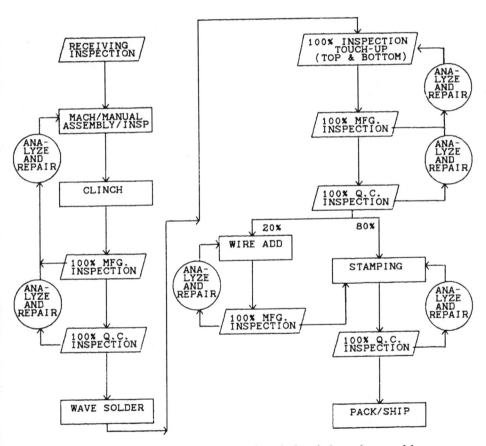

Figure 3.1   Process flow diagram printed circuit board assembly.

provement in both productivity and quality. The tremendous amount of paperwork at each station impacted labor productivity, and a work simplification technique was used that reduced twenty-four forms to five. The bottleneck problems between operations were resolved through the analysis of production time, queuing time, arrival time, and waiting time for jobs. This enabled production line balancing. Employee total involvement and commitment to quality improvement at the source of production enabled production errors to be determined and resolved in real time. The hours and cost of inspection were significantly improved, as shown in Figures 3.2 and 3.3, respectively.

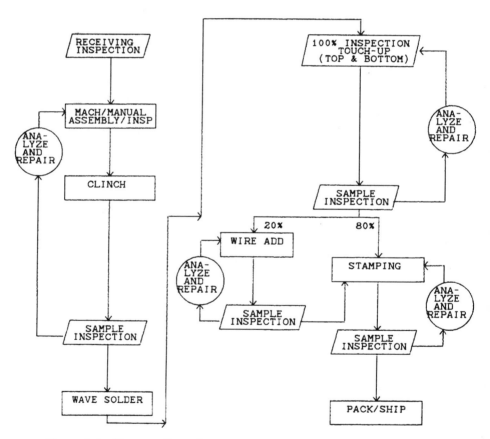

Figure 3.2   Process flow diagram for improved printed circuit board assembly.

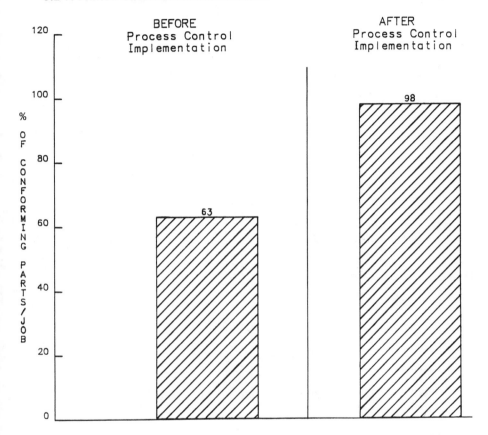

Figure 3.3   First pass yield for the printed circuit board assembly process.

The first-pass yield improved by 35%, as shown in Figure 3.4. There was also an increase in labor and total productivities, as shown in Figures 3.5 and 3.6.

This case study demonstrates that productivity and quality are connected. Techniques implemented to improve quality also improved productivity, and vice versa. There is a correlation between productivity and quality; it is virtually impossible to have one without the other. However, there has to be a balance between the desire to obtain gains in productivity and the desire to improve quality. The productivity and quality objectives, actions, and resources utilized for improvements

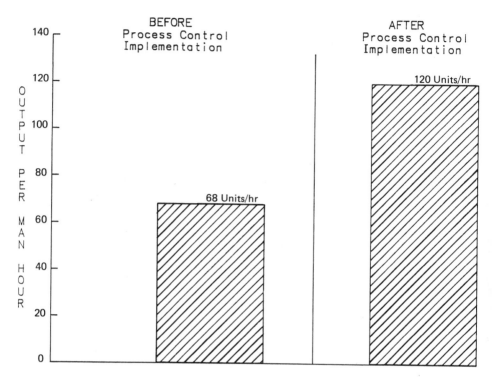

Figure 3.4   Output per labor hour for the printed circuit board assembly process.

must be carefully analyzed. Productivity and quality issues are corrected, interrelated, and inseparable. Productivity increases as quality improves.

3.3   CASE STUDY: A SHOE-MANUFACTURING ENVIRONMENT

A medium-sized shoe-manufacturing company had problems in maintaining a balance of production cost, schedule, and quality. Customer complaints revealed dissatisfaction with sport shoes quality and also delays in receiving orders on time. Dress shoes were an average of 30 days behind the delivery schedule. In order to meet customer demand,

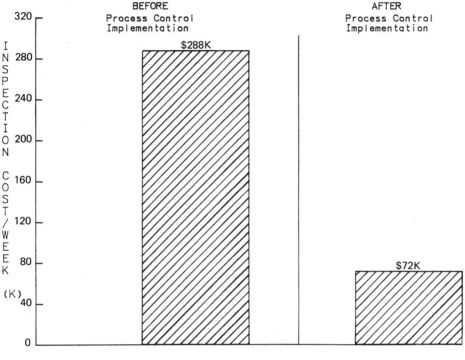

Figure 3.5 Inspection cost per week for the printed circuit board assembly process.

the company placed emphasis on pushing the schedule. This resulted in a high production cost and decreased quality of the shoes produced. A consultant investigated the root cause of the various problems. It was determined that the production process was imbalanced. The input and output of each operation varied significantly. It was also found that the shoe injection molding operation was introducing defects into the product. This caused additional rework to be performed. More inspectors were added to the line to inspect quality into the product. The consultant performed a motion and time study of each operation and recommended an approach to balance the production line. The injection molding operation was redesigned. The results obtained in this study are presented in Table 3.1. From these results, the production

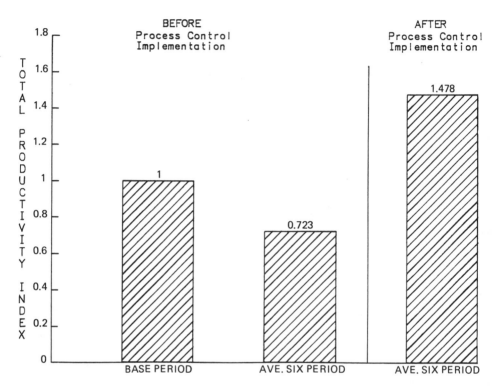

Figure 3.6 Total productivity index for the printed circuit board process.

manager was convinced that a balance can be maintained between production cost, schedule, and quality. It was demonstrated that techniques implemented to improve productivity also have a positive impact on quality and other production variables.

3.4   CASE STUDY: A RESTAURANT

A fast-food restaurant unit was unsure of how to deal with the numbers of complaints received from customers who purchased hamburgers, fish sandwiches, chicken sandwiches, apple pie, french fries, and

Table 3.1　Productivity and Quality Improvement in a Shoe-Manufacturing Company

Variables	Before consultant analysis	After consultant analysis and implementation of new ideas
Output per labor hour (labor productivity)	11	27
Quality defects per unit (%)	7.9	1.4
Work-in-process inventory ($)	$360,000	$57,000
Average cycle time for all shoes produced	17 days	3 days

drinks for quick consumption. Within a period of 3 months there were 506 complaints from customers. The complaints ranged from an improper amount of ketchup on hamburgers to total omission of lettuce and tomatoes on some orders. A total of 216 customers also complained about the amount of time spent waiting on the line to give orders and receive food. A careful analysis of the current mode of operation revealed that the sandwich production process was not streamlined. There were cross motions between work stations, and the arrangement of lettuce, tomatoes, ketchup, salt and pepper, and other cooking ingredients was a scattered process. An attempt was made to understand the number of people expected within a given hour. From records and a random observation of the arrival patterns of customers to the restaurant, it was clear that a new strategy was needed to accommodate peak periods such as lunch from 11:30 a.m. to 1:15 p.m. and dinner from 5:30 to 7:28 p.m. The restaurant implemented two new techniques: cooking ingredients were arranged by sandwich types, and common ingredients were placed in sequence steps that allowed a better flow of work. Cross motion between attendants was eliminated. The chances of making an error were minimized. The queuing technique was used to predict the arrival pattern and total number of customers expected within a given time period. Implementation of these two new ideas resulted in a 79% reduction in the total number of customer complaints within an equivalent 3-month period. The average waiting time in the line was reduced from 11 minutes to 4 minutes during the peak demand periods. The fast-food unit was able to attract additional customers.

This case study shows that pressure on productivity or output in one direction without careful attention to quality can be costly. However, the improvement in the quality of services in this case study was found to have a positive impact on productivity and the overall profit posture.

3.5 CASE STUDY: A BANK

A medium-sized bank had received a series of complaints from checking account customers. At the end of a month, 921 customers complained of having not received the correct canceled check, or the wrong transaction number was recorded in their statements. The check processing operation was a manual method involving almost thirty operators. The operators utilized serial code devices and organized the reader work area to process each customer batch of checks from the main input source. A careful analysis of the operation revealed that human error and the volume of checks processed were major contributors to the customer complaints. It was recommended that the management of the bank implement a check processing serialized conveyor. The machine had the capability of organizing checks by serial number, last name, and address code. After the implementation of the new system, the jobs of thirty operators were eliminated. The number of complaints from the checking account customers was reduced from 921 to 63 per month. Most complaints after the implementation of the new check serialized conveyor system were about mechanical damage from the system. These complaints were kept at a minimum through proper overhauling of the system.

As shown in these four case studies, productivity and quality objectives, issues, problems, solutions, and results are interrelated. It is almost impossible to have one without the other. Organizations that put a program in place to manage productivity and quality together are bound to receive positive results. However, benefit also comes from the application of creativity from employees and management in addressing various day-to-day production or service challenges.

3.6 UNDERSTANDING THE BASIC COST OF QUALITY

In Chapter 1, the four categories of quality costs were discussed. In this section, an approach is presented on how the basic cost of quality is computed. The costs of conformance are made up of costs associated with prevention and appraisal. The costs of nonconformance are made up of internal and external failure costs.

Let

CA_{ijt} = appraisal costs incurred in performing task i, in site j, in period t, resulting from inspections and repeated checks

CP_{ijt} = prevention costs incurred in performing task i, in site j, in period t, resulting from preventive action taken to avoid defects in design, development, manufacturing, service, and purchasing

CI_{ijt} = internal failure costs incurred in performing task i, in site j, in period t, resulting from items that do not conform to requirements or specifications prior to shipment from the activity or site; i = 1, 2, 3, ..., m; j = 1, 2, 3, ..., k; and t = 1, 2, ..., n

CE_{ijt} = external failure costs incurred in performing task i, in site j, in period t, resulting from items that do not conform to requirements after shipments from site

TC_{ijt} = total cost of quality incurred in performing task i, in site j, in period t

AV_{ijt} = accepted units of measure for cost of quality of task i, in site j, in period t (AV can be in hours, weights, or monetary value)

$$TC_{ijt} = (CA_{ijt} + CP_{ijt} + CI_{ijt} + CE_{ijt}) \qquad (3.1)$$

where TC_{ijt} is expressed in hours of work or monetary terms.

The following steps can be used to obtain TC_{ijt}.

1. Choose a specific, observable task.
2. List each element involved in the task.
3. Identify the input and output of the task.
4. Identify the effort spent on all the variables specified in expression (3.1). The components are

CA_{ijt}, CP_{ijt}, CI_{ijt}, and CE_{ijt}.

5. Estimate in quantitative terms the time (hours) spent for each item in step 4, and compute TC_{ij} in monetary terms by conversion.

Using the values of TC_{ij} obtained, the analyst should investigate contributing factors to each cost category and interact with both supplier and customer or process owner to provide actions that eliminate or minimize the various cost categories. In performing the analysis, the analyst must always recognize the relationship that exists between supplier, process owner, and customer, as shown in Figure 3.7. Work forms for the computation of costs of quality are presented in Appendix D.

3.7 COST OF INSPECTION PER SAMPLE

In a given task, the cost of inspection per sample can be computed by expression (3.2):

$$CI_{ijt} = C_f + NC_v + (CI)(PI) \tag{3.2}$$

where:

CI_{ijt} = cost of inspection per sample in task i, in site j, in period t

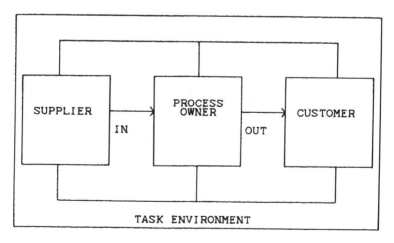

Figure 3.7 Relationship between supplier, process owner, and customer in a task environment.

PI = probability of inferring the process has changed when it has not

C_f = fixed cost of inspection per sample

N = sample size

C_v = variable cost of inspection per unit

CI = cost of inferring that the process has changed when it has not

The conditional loss per inspection period can be determined by expression (3.3):

$$CL = D_c \left(\frac{R}{N_s}\right)^K_{K=2} \frac{C_k}{N_2} (C_k - C_1) \left(\frac{1}{1 - P_k} - 0.5\right) \qquad (3.3)$$

where:

CL = conditional loss per inspection period

f_k = fraction defective in state k

K = states of system (K = 1, 2, ..., M), with state 1 the desired state

N_s = number of samples per period

C_k = expected number of changes to state k in period

P_k = probability that a given plan will fail to detect a change to state k

R = production rate per period

D_c = cost of a defect

Duncan (1965), Fetter (1967), and ASTM (1951) provide a more theoretical detailed discussion of sampling techniques and costs.

The next chapter presents a formalized approach for measuring productivity. Emphasis is placed on total productivity measurement. The productivity measurement models presented can be used at all levels of the productivity and quality management hierarchy (PQMH).

REFERENCES

American Society for Testing and Materials, 1951. *ASTM Manual on Quality Control of Materials*, Philadelphia, Pennsylvania.

Cole, R. E., 1981. The Japanese lesson in quality. *Technology Review*, July, p. 29.

Danforth, D. D., 1984. Quality means doing the job right the first time. *The Wall Street Journal*, March 21, p. 33.

Duncan, A. J., 1965. *Quality Control and Industrial Statistics*. Richard D. Irwin, Homewood, Illinois.

Feigenbaum, A. V., 1979. American manufacturers strive for quality—Japanese style. *Business Week*, March 12, p. 5.

Fetter, R. B., 1967. *The Quality Control System*. Richard D. Irwin, Homewood, Illinois.

Midas, M. T., 1981. The productivity-quality connection. *Design News*, December 7, p. 56.

4

Productivity Measurement

The concept of a formal approach to productivity measurement at the basic level in companies and organizations is still in its infant stages of conceptualization and rationalization. This chapter presents the benefits of productivity measurement, a summary of the four most commonly used productivity indices in companies, and a critique of their effectiveness. Both the task-oriented and the technology-oriented total productivity measurement (TOTP) model, which incorporates a new method of inputs and outputs in the work environment, are presented. Case studies are offered to show the validity of the models.

4.1 BENEFITS OF PRODUCTIVITY MEASUREMENT IN COMPANIES AND ORGANIZATIONS

Edosomwan (1985a, p. 1) pointed out that in order for companies to effectively compete in the world market and contribute to the national growth rate of productivity in both the short run and the long run, it is necessary for them to institute a formal productivity measurement system. Such a system can have the following important benefits.

1. Productivity measurement provides an important motivation for better performance, since it helps to identify on what basis the individual task, project, product, or customer is to be measured. It provides the basis for planning the profit level in a company.
2. Productivity measurement highlights by means of indices those areas within the company that have potential improvement possibilities. Productivity values and indices also provide a way of detecting deviations from established standards on a timely basis so that something is done about such deviations.
3. Productivity measurement creates a basis for the effective supervision of necessary actions to be taken and improves decision-making through better understanding of the effect of actions already taken to address a given problem.
4. Productivity measurement can be used to compare the performance levels of individuals, work groups, tasks, projects, departments, and firms as a whole.
5. Productivity measurement facilitates better resource planning and projections in both the short and the long run. It also simplifies communication by providing common measures, language, and concepts with which to think, talk, and evaluate the business in quantitative terms.

4.2 DIFFICULTIES IN MEASURING PRODUCTIVITY

Kendrick (1984, p. 18) pointed out that the operational concept of productivity involves many detailed definitions and statistical problems, including

1. Measuring outputs whose characteristics may change over time
2. Defining and measuring real capital stocks and inputs as well as labor inputs when the characteristics of both factors are diverse and changing
3. Aggregating heterogeneous units of output and input; Kendrick further pointed out that these problems would exist even if data were perfect and suggests using prices or unit costs for aggregation purposes

4.3 BRIEF DESCRIPTION OF THE MOST COMMONLY USED PRODUCTIVITY MEASUREMENT APPROACHES IN COMPANIES AND A CRITIQUE OF THEIR EFFECTIVENESS

Economic studies reported in the literature so far have used four types of productivity measures.

4.3.1 Partial Productivity

Partial productivity is the ratio of output to one class of input. Output per labor hour is the best example of a partial productivity measure and is the one most commonly used. Most productivity indices published by the U.S. Department of Labor (1980) are partial measures. Melman (1956), Mundel (1976), and Turner (1980) have also used such measures. However, there is a danger in using partial measures of productivity. Siegel (1976), Craig and Harris (1972, 1973), Sumanth (1979), and Edosomwan (1980) pointed out that a partial measure of productivity could be misleading when viewed alone. For example, a high material productivity could project that a company is doing well although indeed, capital productivity, energy productivity, labor productivity, and other indices may be low. The actual danger of partial measure is that it overemphasizes one input and others are neglected. Turner (1980) also agreed that a partial measure of productivity, such as output per labor hour, could not be interpreted as an overall productivity measure since it does not take into account all input cost. Kendrick (1984) pointed out that partial productivity, such as labor productivity, reflects only one input measure and cannot be viewed as overall productivity since it reflects factor substitutions. The author recommends total productivity for overall productivity measurement and total factor productivity for when labor and capital inputs are considered. The survey reported by Sumanth (1981) also revealed that a problem still exists in the lack of attention to total productivity measures. As shown in Tables 4.1 and 4.2 and Figure 4.1, the survey indicates that 91.7% of engineering organizations in industrial companies use nonstandard productivity indicators, as do 70% in nonindustrial companies.

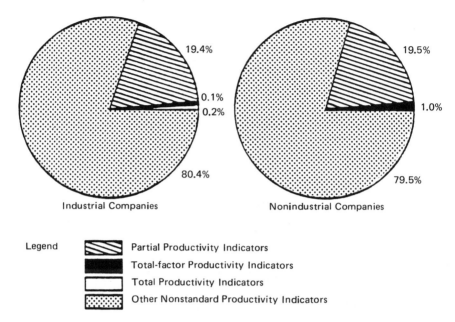

Figure 4.1   Productivity indicators. (*Source*: David J. Sumanth, Productivity indicators used by major U.S. manufacturing companies: The results of a survey. *Industrial Engineering*, May 1981, p. 73; and David J. Sumanth, Survey results: How major non-industrial corporations measure their productivity. *Industrial Engineering*, September 1981, p. 32. Reprinted with Permission.)

4.3.2   Utility Index

This approach is one of the techniques proposed to measure the productivity of a firm. Stewart (1978) presented the utility concept for measuring manufacturing productivity. Stewart defines productivity as "the ratio of performance toward organizational objectives to the totality of input parameter." The essence of Stewart's utility approach is to combine "surrogate measures" to produce a single number. He defines a surrogate measure as "one which is used in place of another and is often used when the desired measure is unobtainable." One weakness of this index is that it fails to attribute performance to the various individual inputs, such as raw material, technology, and en-

Table 4.1 Productivity Indicators in Industrial Companies

Function	Partial productivity indicators		Total-factor productivity indicators		Total productivity indicators		Other nonstandard productivity indicators	
	No. of times reported	%	No. of times reported	%	No. of times reported	%	No. of times reported	%
Manufacturing	30	41.1	1	1.4	2	2.7	40	54.8
Sales	16	39.0	0	0.0	0	0.0	25	61.0
Marketing	6	22.2	0	0.0	0	0.0	21	77.8
Purchasing	4	15.4	0	0.0	0	0.0	22	84.6
Personnel	3	11.5	0	0.0	0	0.0	23	88.5
Finance and accounting	5	23.8	0	0.0	0	0.0	16	76.2
Legal	1	14.3	0	0.0	0	0.0	6	85.7
Engineering	2	8.3	0	0.0	0	0.0	22	91.7
Research and development	0	0.0	0	0.0	0	0.0	16	100.0
Quality assurance	7	18.9	0	0.0	0	0.0	30	81.1
Maintenance	5	16.1	0	0.0	0	0.0	26	83.9
Industrial engineering	0	0.0	0	0.0	0	0.0	18	100.0
Data processing	2	7.7	0	0.0	0	0.0	24	92.3
Administration	4	18.2	0	0.0	0	0.0	18	81.8
Word processing	10	55.6	0	0.0	0	0.0	8	44.4
Distribution and warehousing	7	24.1	0	0.0	0	0.0	22	75.9
Planning	1	12.5	0	0.0	0	0.0	7	87.5
Average (%)		19.3		0.1		0.2		80.4

Source: David J. Sumanth, Productivity indicators used by major U.S. manufacturing companies: The results of a survey. Industrial Engineering, May 1981, p. 73. Reprinted with permission.

Table 4.2 Productivity Indicators in Nonindustrial Companies

Function	Partial productivity indicators		Total-factor productivity indicators		Total productivity indicators		Other nonstandard productivity indicators	
	No. of times reported	%	No. of times reported	%	No. of times reported	%	No. of times reported	%
Manufacturing (operations)	9	36.0	1	4.0	0	0.0	15	60.0
Sales	12	42.9	1	3.6	0	0.0	15	53.5
Marketing	1	9.1	1	9.1	0	0.0	9	81.8
Purchasing	0	0.0	0	0.0	0	0.0	11	100.0
Personnel	1	8.3	0	0.0	0	0.0	11	91.7
Finance and accounting	5	31.3	0	0.0	0	0.0	11	68.7
Legal	0	0.0	0	0.0	0	0.0	4	100.0
Engineering	3	30.0	0	0.0	0	0.0	7	70.0
Research and development	0	0.0	0	0.0	0	0.0	4	100.0
Quality assurance	0	0.0	0	0.0	0	0.0	5	100.0
Maintenance	2	18.2	0	0.0	0	0.0	9	81.8
Industrial engineering	1	10.0	0	0.0	0	0.0	9	90.0
Data processing	1	6.7	0	0.0	0	0.0	14	93.3
Administration	2	20.0	0	0.0	0	0.0	8	80.0
Word processing	8	53.3	0	0.0	0	0.0	7	46.7
Distribution and warehousing	7	53.8	0	0.0	0	0.0	6	46.2
Planning	1	12.5	0	0.0	0	0.0	7	87.5
Average (%)		19.5		1.0		0.0		79.5

Source: David J. Sumanth, Survey results: How major non-industrial corporations measure their productivity. *Industrial Engineering,* September 1981, p. 32. Reprinted with permission.

ergy. The utility index offers little or no help in pinpointing specific inputs as possible sources for improving the firm's productivity.

4.3.3 Total-Factor Productivity

Total-factor productivity is defined as "the ratio of net output to the sum of associated labor and capital (factor inputs)." One such measure has been recommended by Mali (1978), Kendrick and Creamer (1965), and Taylor and Davis (1977). One disadvantage of this measure is that it omits the cost of materials, which is one of the vital inputs in business, from the denominator, although it is subtracted from the gross output. Omitting such items as raw materials, supplies, and purchase parts can make studies of cost-price relationships difficult, especially at the firm level. As with other inputs, material and computer inputs form important cost elements, and any savings obtained through their usage per unit of output affects the total unit cost of output and hence the prices.

4.3.4 Total Productivity Measure

The total productivity measure considers total output in relation to total input and has been proposed by most authors. For the most part there have been significant variations in the definition of the input and output elements. Various authors have also proposed different allocation criteria for specifying the proportional contributions of each input element to the final output.

Kendrick and Creamer (1965), Craig and Harris (1972, 1973), Hines (1976) Mali (1978), Sumanth (1979), and Kendrick (1984) have proposed such measures. Some total productivity measures are presented vaguely and as general relationships between some measurable input and output elements; others use unquantifiable descriptions. For example, Mali (1978) defined productivity as the ratio of effectiveness to efficiency. It might be difficult to quantify effectiveness numerically in a real-world application.

Other measures of productivity proposed in the literature include the array approach of Dewitt (1970, 1976), the financial ratio approach of Tucker (1961), Gold (1976), and Aggrawal (1979); the servosystem approach of Hershauer and Ruch (1978); the capital budget approach of Mao (1965); and the production function approach of Dhrymes (1963).

In the various productivity measures outlined above, only a few total productivity measurement models focus on both the production and service environment and without recognition of the impact of computers and robotics on productivity results. In most cases computer and robot applications are treated as fixed capital, or only the supplies needed to operate the computer and robots are treated as an information processing cost.

The current available theory of productivity is not concerned with productivity measurement at detailed functional levels, such as by task or by project. Nothing has been done in developing useful relationships between the total and partial productivities of multitasks and no attempts made to relate total and partial productivity indices and values to a costing philosophy adopted by a given firm.

Methodologically, aggregation techniques have been used to measure the overall productivity of a firm. At the firm level, such aggregation techniques have provided different results depending on the type of allocation criteria used for other individual resources, overhead cost, the definition of fixed and working capital, floor space, and other factors.

The available productivity measurement models also omit the new technological version in which machinery becomes a variable cost. However, some of the total productivity measures, such as those of Kendrick and Creamer (1965), Sumanth (1979), and Craig and Harris (1972, 1973), provide a basis for quantifying in meaningful terms some of the measurable input and output elements. From the discussion presented so far in this section, the shortcomings and strengths of the present productivity measurement approaches at the firm level have been highlighted. Obviously, there is a need for appropriate productivity measurement by task and technology, for example, and through aggregation techniques to extend such measures to the firm level.

4.4   THE TASK-ORIENTED TOTAL PRODUCTIVITY MEASUREMENT (TOTP) MODEL

The task-oriented total productivity model developed and recommended by Edosomwan (1985a) is based on all possible measurable output and input components. An incremental analysis is somewhat implicit in the model. The measures derived from this model are in the form of an index that intuitively has the following properties and advantages.

1. The indices derived use the broadest possible input (labor, materials, energy, robotics, computers, capital, data processing, and other administrative expenses) and output (finished units produced, partial units produced, and other output associated with units produced).
2. The productivity indices derived vary with changes in task parameters, resources utilized, and output obtained from the transformation of resources.
3. The productivity indices derived are comparable over time and can objectively be used to measure the productivity of tasks, customs, products, projects, work groups, departments, division, and company.
4. They provide a means of focusing on key problem areas for productivity improvement. The indices identify which particular input resources are utilized inefficiently so that an improvement action plan can be implemented.
5. The indices can be used in productivity planning and improvement phases. They also offer a basis for companies in planning every phase of a product or technology development cycle.

4.4.1 Key Definitions Associated with the Task-Oriented Total Productivity Measurement Model

Task: At the basic level, a task is a unit of work accomplished primarily at a single location (site), by a single agent, during a single time period, producing useful output from some resources available.

Total productivity is the ratio of total measurable output (total finished units produced, partial units produced, and other output associated with units produced) to the sum of all the measurable inputs (labor, materials, capital, energy, robotics, computers, data processing, and other administrative expenses) utilized for production.

Total factor productivity is the ratio of total measurable output to the sum of labor and capital inputs.

Partial productivity is the ratio of total measurable output to one class of measurable input (for example, labor hours utilized for production or service).

4.4.2 Input and Output Components of the Task-Oriented Total Productivity Measurement Model

The input and output components of the task-oriented total productivity measurement model are shown schematically in Figures 4.2 and 4.3. (A description of each component of the model is offered in Section 4.4).

4.4.3 Derivation of Productivity Values and Index Notations

Let

i = manufacturing or service task

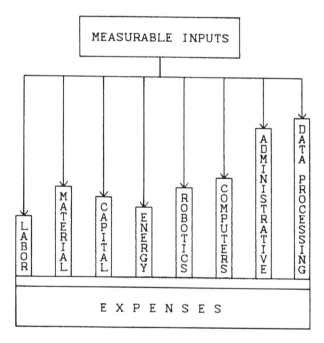

Figure 4.2   Input components considered in the task-oriented total productivity model. (From Edosomwan, 1985a and 1986.)

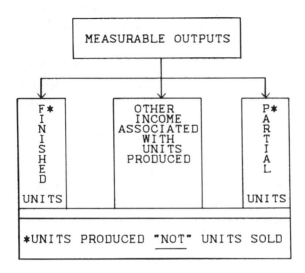

Figure 4.3   Output components considered in the task-oriented total productivity model. (From Edosomwan, 1985a and 1986.)

j = manufacturing or service plant location (j = 1, 2, 3, ..., m) (model can be applied simultaneously in more than one site)

t = study period t = 1, 2, 3, ..., n

O_{ijt} = total quantity of units produced (total output) by task i, in site j, in period t

F_{ijt} = finished task (finished units produced) by task i, in site j, in period t

PC_{ijt} = percentage completion of partial units produced by task i, in site j, in period t

P_{ijt} = partially completed task (partial units of units produced) by task i, in site j, in period it

SP_{ijt} = base period selling price per unit for a unit produced by task i, in site j, in period t

CR_{ijt} = variable computer-related expense input in dollars utilized by task i, in site j, in period t

RR_{ijt} = variable robotics-related expense input in dollars utilized by task i, in site j, in period t

L_{ijt} = labor hours input utilized by task i, in site j, in period t

LR_{ijt} = labor rate per hour utilized by task i, in site j, in period t

M_{ijt} = material and purchase parts expense input in dollars utilized by task i, in site j, in period t

C_{ijt} = capital-related expense (includes fixed and working capital, such as cash, accounts receivable, tools, plant, buildings, amortization, and research and development) input in dollars utilized by task i, in site j, in period t (capital is computed using the lease value concept)

E_{ijt} = energy-related expense (includes electricity, solar energy, water, coal, and gas) input in dollars utilized by task i, in site j, in period t

DP_{ijt} = variable data processing expense input in dollars utilized by task i, in site j, in period t

OE_{ijt} = other administrative expense-related expense input in dollars utilized by task i, in site j, in period t (other expense includes travel, taxes, professional fees, marketing, research and development, and general administration)

Y_{ijt} = total productivity of task i, in site j, in period t

TF_{ijt} = total factor productivity of task i, in site j, in period t

P_{ijt} = partial productivity of task i with respect to one input in site j, in period t

I_{ijt} = total input of task i, in site j, in period t

T_b = base period time (a reference period to which the output and input in monetary terms is reduced; thus, the total productivity of a task expressed as dollar output per dollar input in constant dollars)

The total and partial productivities for a given task expressed as follows. Let Y_{ijt} = total productivity of task i, performed in site j, in period t.

$$Y_{ijt} = \frac{\text{total measurable output of task i, performed in site j, in period t}}{\text{total measurable input of task i, performed in site j, in period t}} \qquad (4.1)$$

$$Y_{ijt} = \frac{O_{ijt}}{I_{ijt}} \tag{4.2}$$

$$Y_{ijt} = \frac{(F_{ijt})(SP_{ijt}) + (P_{ijt})(SP_{ijt})(PC_{ijt})}{(CR_{ijt}) + (L_{ijt})(LR_{ijt}) + (M_{ijt}) + (E_{ijt}) + (OE_{ijt})} \tag{4.3}$$
$$+ (C_{ijt}) + (DP_{ijt})$$

Let

P_{ijt} = partial productivity of task i, with respect to labor input in site j, in period t.

$$P_{ijt} = \frac{\text{total measurable output of task i, performed in site j, in period t}}{\text{measurable labor input of task i, performed in site j, in period t}} \tag{4.4}$$

$$P_{ijt} = \frac{(F_{ijt})(SP_{ijt}) + (P_{ijt})(SP_{ijt})(PC_{ijt})}{(L_{ijt})(LR_{ijt})} \tag{4.5}$$

The total, total-factor, and partial productivities for a given task in any period are schematically presented in Figure 4.4.

4.4.4 Implementation Methodology for the Task-Oriented Total Productivity Measurement Model

Implementing the task-oriented total productivity measurement model in a company or organization is relatively easy if the step-by-step procedure presented schematically in Figure 4.5 is followed. Each step will now be described.

Step 1: Familiarization Sessions

In this step, a formal study of the work environment and processes, procedures, cost accounting system, types of product produced or services, types of operations, and tasks performed is carefully conducted. A formal acquaintance with key personnel in the various departments is essential. In order to ensure total involvement of everyone in the organization, a kickoff meeting is recommended. Such a meeting must have the support of upper management. The meeting should be geared toward explaining the objectives and benefits of the productivity measurement program to every level of employee and management.

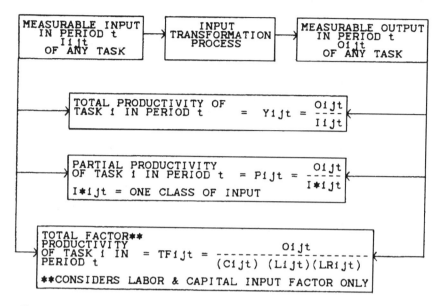

Figure 4.4   Total, total factor, and partial productivities for a given task in any site j in period t.

The formal meeting should establish commitments from everyone and provide a good communication link among people and departments. The elements of the familiarization circle are shown in Figure 4.6.

Step 2: Task Significance, Input-Output Analysis

There are two approaches in trying to select the scope of productivity measurement within an organization. One approach, particular to small firms, involves a choice of measuring the productivity of every task, department, product, customer, and the firm as whole. The other approach, particular to larger organizations, is to measure the productivity of tasks that account for a significant amount of business volume and revenue. In this step the productivity coordinator, working with the team members of the measurement program, selects a set of tasks by using historical data on sales, profits, and costs. It is recommended that the sales records and income statements for the last 24-month account-

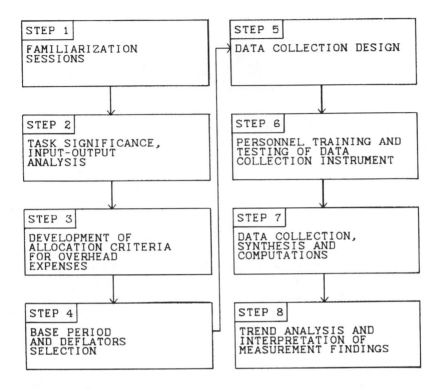

Figure 4.5   Implementation steps for the task-oriented total productivity measurement model.

ing period be used to determine the percentage of contribution for each task performed. Applying Pareto's law, determine and select those tasks that account for 80 percent or more of the company's total sales and cost. For an organization starting business for the first time and also instituting a productivity measurement program for the first time, select the tasks by estimating the products or services that account for 80 percent or more of the organization's total production and sales. The final selection of tasks to be considered in the productivity measurement program should also be supplement by managerial judgment. The elements of the task significance and input-output analysis are shown schematically in Figure 4.7.

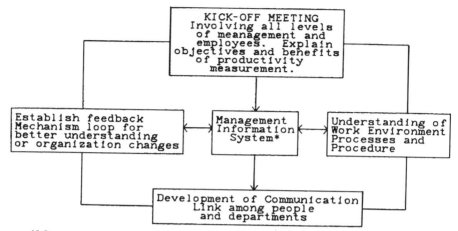

*Management Information System is a database containing all information on productivity and quality issues and other vital information on business operations within the organization.

Figure 4.6 Components of productivity measurement familiarization sessions.

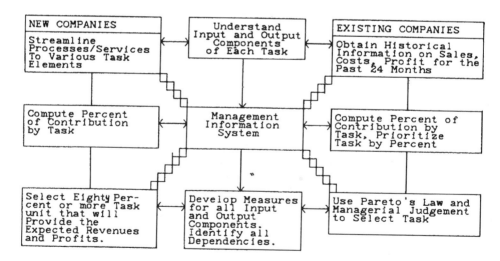

Figure 4.7 Task significance, input-output analysis for the productivity measurement program.

Step 3: Development of Allocation Criteria for Overhead Expenses

Although several overhead allocation criteria exist, the most generally used traditional cost accounting and industrial engineering approach is the proportional contribution of direct hours to allocate overhead expense to a product or task. This approach requires daily direct labor recording and the derivation of burden rate from net expense and direct hours. Therefore, product costs included labor and burden. Allocation of overhead is done using direct hours and one burden rate applied to all products produced or services rendered by the organization. In direct hours allocation criteria, some product costs may be misallocated, especially in situations in which multiple products are produced. It must also be pointed out that the allocation criteria for the output and input elements needed for productivity measurement may vary depending on the type of production or service environment, costs, accounting information, and managerial preferences. Ideally, it is more convenient to capture all output and input elements for each task. Streamlining input and output components is difficult and uneconomical if the right criteria are not used. Using the task-oriented total productivity measurement model, two types of allocation schemes for input and output components are recommended.

The first allocation scheme is based on the concept of "proportion." The proportional contribution is sales, profit, cost, and quantity produced is used as a factor to allocate overhead expenses to tasks. The second allocation scheme is based on the concept of "complexity factor." This scheme proposes the use of a product complexity factor in allocating overhead expenses to tasks. The complexity factor is determined through such items as component counts, insertion rates, component preparation time, and design configuration. Energy usage, floor space, telephone, indirect efforts, and supervision costs are seen as a function of the complexity factor. The elements of the allocation criteria steps are shown schematically in Figure 4.8.

Step 4: Base Period and Deflator Selection

The base period is a reference period in which the output and input values in monetary terms are reduced. Thus, the total productivity of a task is expressed as dollar output per dollar input in constant dollars. The criteria for selecting a base period depends upon

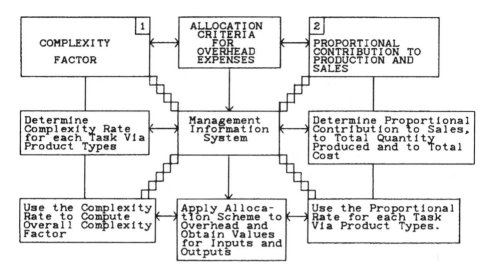

Figure 4.8 Elements in the development of allocation criteria for overhead expenses in productivity measurement.

1. Whether a productivity measurement program (PMP) is instituted for the first time
2. How often new tasks are introduced
3. How often the products produced or services rendered are changed
4. Whether the productivity values and indices are established on a monthly, quarterly, semiannual, or annual basis
5. Whether there have been abnormal developments within the organization, such as unprecedented layoffs, strikes, and lockouts
6. How often the data collection system is updated
7. Whether there has been any abnormal reorganization within the organization, such as conversion of direct effort labor to indirect, or vice versa
8. Whether there have been seasonal demand patterns for the product produced or services offered by the organization
9. Whether the trend analysis time period chosen for the management reporting system includes or excludes a particular business time frame based on managerial judgment

10. Whether there is significant effect of other external factors, such as interest rates, on the organization's input and output levels

The elements involved in the selection of the base period are shown schematically in Figure 4.9. Deflators are needed in situations in which the output and input cannot be expressed in physical units. The main sources of deflators are consumer price indices, producer price indices, wage rates, material price indices, energy price indices, and the like. These indices are obtainable from the monthly labor reviews published by the Bureau of Labor Statistics, Washington, D.C. In situations in which it is difficult to obtain the deflators on real-time basis, they can be forecast using historical data from the price indices. Examples of forecasted deflators for labor, energy, capital, administrative expenses, materials, and output components will now be presented. The consumer price indices obtained from the monthly labor review are used. Regression analysis is used for forecasting deflators and indices for future time periods. Whether the forecasted values had a high correlation was also determined.

Labor Input

From Table 4.3, let the total number of periods in the regression analysis n = 13 (see also Table 4.4 and 4.5).

$$b = \frac{nXY - (\Delta X\ \Delta Y)}{nX^2 - (\Delta X)^2}$$

$$= \frac{(13 \times 385.94) - (91 \times 54.62)}{(13 \times 819) - 91^2}$$

$$= 0.0198$$

$$a = \frac{Y - bX}{n}$$

$$= \frac{54.62 - (0.0198 \times 91)}{13}$$

$$= 4.06$$

$$Y = a + bx$$

$$= 4.06 + 0.0198X$$

Table 4.3   Labor Input Regression Analysis Using CPI

Month X	Index Y	XY	X^2	Y^2
1 Nov. 1978	4.04	4.04	1	16.32
2 Dec. 1978	4.08	8.16	4	16.65
3 Jan. 1979	4.17	12.51	9	17.39
4 Feb. 1979	4.17	16.68	16	17.39
5 Mar. 1979	4.19	20.95	25	17.56
6 Apr. 1979	4.19	25.14	36	17.56
7 May 1979	4.20	29.40	49	17.64
8 June 1979	4.21	33.68	64	17.72
9 July 1979	4.23	38.07	81	17.89
10 Aug. 1979	4.21	42.1	100	17.72
11 Sept. 1979	4.28	47.08	121	18.32
12 Oct. 1979	4.32	51.84	144	18.66
13 Nov. 1979	4.33	56.29	169	18.75
X = 91	Y = 54.62	XY = 385.94	X^2 = 819	Y^2 = 229.57

Note: CPI indexes were obtained from the *Monthly Labor Review.*

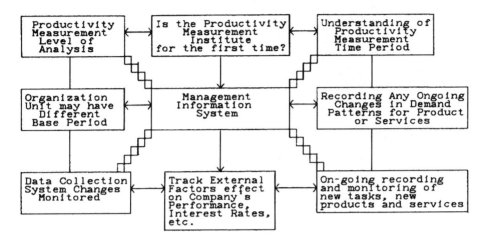

Figure 4.9   Elements in base period selection.

Table 4.4   Predicted Index Values for Labor Input

Month X		Equation	Index Y
14	Dec. 1979	Y = 4.06 + 0.0198 (14)	4.34
15	Jan. 1980	Y = 4.06 + 0.0198 (15)	4.36
16	Feb. 1980	Y = 4.06 + 0.0198 (16)	4.38
17	Mar. 1980	Y = 4.06 + 0.0198 (17)	4.40
18	Apr. 1980	Y = 4.06 + 0.0198 (18)	4.42
19	May 1980	Y = 4.06 + 0.0198 (19)	4.44
20	June 1980	Y = 4.06 + 0.0198 (20)	4.46

$$r = \frac{nXY - (\Delta X \, \Delta Y)}{\sqrt{nX^2 - (\Delta X)^2} \sqrt{nY^2 - (\Delta Y)^2}}$$

$$= \frac{46.8}{\sqrt{2366} \sqrt{(13 \times 299.57) - 54.62)^2}}$$

$$= 0.93 \quad \textit{Very high correlation}$$

Energy Input

From Table 4.6, let n = 12 (see also Table 7 and 8).

$$b = \frac{nXY - (\Delta X \, \Delta Y)}{nX^2 - X^2}$$

$$= \frac{(12 \times 21779.3) - (78 \times 3159.2)}{(12 \times 650) - 78)^2}$$

$$= 8.70$$

$$a = \frac{Y - bX}{n}$$

$$= \frac{3159.2 - (8.70 \times 78)}{12}$$

$$= 206.7$$

$$Y = a + bx$$

$$= 206.7 + 8.70X$$

$$r = \frac{nXY - (\Delta X \, \Delta Y)}{\sqrt{nX^2 - (\Delta X)^2} \sqrt{nY^2 - (\Delta Y)^2}}$$

$$= \frac{14934}{\sqrt{1716} \sqrt{(12 \times 842940.7) - 3159.2)^2}}$$

$$= 0.98 \quad \textit{Very high correlation}$$

Output, Material, Capital, and Other Administrative Expense Regression Analysis

From Table 4.9, let n = 12 (see also Tables 4.10 and 4.11).

Table 4.5   Deflators for Labor Input

	Jan. 1980	Feb. 1980	Mar. 1980	Apr. 1980	May 1980	June 1980
Index	4.36	4.38	4.40	4.42	4.44	4.46
Deflator	1.000	1.005	1.009	1.013	1.018	1.022

Table 4.6   Energy Input Regression Analysis Using CPI

Month X		Index Y	XY	X^2	Y^2
1	Nov. 1978	226.0	226.0	1	51,076.0
2	Dec. 1978	228.5	457.0	4	52,212.2
3	Jan. 1979	231.8	695.4	9	53,731.2
4	Feb. 1979	235.3	941.2	16	55,366.1
5	Mar. 1979	241.7	1208.5	25	58,418.9
6	Apr. 1979	251.2	1507.2	36	63,101.4
7	May 1979	262.2	1835.4	49	68,748.8
8	June 1979	277.3	2218.4	64	76,895.3
9	July 1979	289.2	2602.8	81	83,636.4
10	Aug. 1979	298.8	2988.0	100	89,281.4
11	Sept. 1979	307.0	3377.0	121	94,249.0
12	Oct. 1979	310.2	3722.4	144	96,224.0
$X = 78$		$Y = 3159.2$	$XY = 21,779.3$	$X^2 = 650$	$Y^2 = 842,940.7$

$$b = \frac{nXY - (\Delta X \, \Delta Y)}{nX^2 - X^2}$$

$$= \frac{(12 \times 8600.4) - (78 \times 1305.9)}{(12 \times 650) - 78)^2}$$

$$= 0.784$$

Table 4.7 Predicted Index Values for Energy Input

Month X		Equation	Index Y
13	Nov. 1979	Y = 206.7 + 8.70 (13)	319.2
14	Dec. 1979	Y = 206.7 + 8.70 (14)	328.5
15	Jan. 1980	Y = 206.7 + 8.70 (15)	337.2
16	Feb. 1980	Y = 206.7 + 8.70 (16)	345.9
17	Mar. 1980	Y = 206.7 + 8.70 (17)	354.6
18	Apr. 1980	Y = 206.7 + 8.70 (18)	363.3
19	May 1980	Y = 206.7 + 8.70 (19)	372.0
20	June 1980	Y = 206.7 + 8.70 (20)	380.7

Table 4.8 Deflators for Energy Input

	Jan. 1980	Feb. 1980	Mar. 1980	Apr. 1980	May 1980	June 1980
Index	337.2	345.9	354.6	363.3	272.0	380.7
Deflator	1.000	1.026	1.052	1.077	1.103	1.129

$$a = \frac{Y - bX}{n}$$

$$= \frac{1305.9 - (0.784 \times 78)}{12}$$

$$= 103.7$$

Table 4.9   Output, Material, Capital, and Other Administrative
Expenses: Regression Analysis Using Consumer Price Indexes

Month X		Index Y	XY	X^2	Y^2
1	Nov. 1978	105.6	105.6	1	11,151.4
2	Dec. 1978	105.8	211.6	4	11,193.6
3	Jan. 1979	105.4	316.2	9	11,109.2
4	Feb. 1979	105.3	421.2	16	11,088.1
5	Mar. 1979	106.9	534.5	25	11,427.6
6	Apr. 1979	109.2	655.2	36	11,924.6
7	May 1979	109.9	769.3	49	12,078.0
8	June 1979	110.4	883.2	64	12,188.2
9	July 1979	110.9	998.1	81	12,298.8
10	Aug. 1979	110.0	1100.0	100	12,100.0
11	Sept. 1979	112.5	1237.5	121	12,656.2
12	Oct. 1979	114.0	1368.0	144	12,996.0
X = 78		Y = 1305.9	XY = 8600.4	X^2 = 650	Y^2 = 142,211.7

$$Y = a + bx$$

$$= 103.7 + 0.784X$$

$$r = \frac{nXY - XY}{\sqrt{nX^2 - (\Delta X)^2} \sqrt{nY^2 - (\Delta Y)^2}}$$

$$= \frac{1344.6}{\sqrt{1716} \sqrt{(12 \times 142m211.7) - 1305.9)^2}}$$

$$= 0.95 \quad \textit{Very high correlation}$$

The elements involved in deflator selection are shown schematically in Figure 4.10

Step 5: Data Collection Design

In this step, various forms and instruments are designed to capture the input and output elements needed for the task-oriented productivity

Table 4.10   Predicted Index Values for Output, Material, Capital, and Other Expenses

Month X		Equation	Index Y
13	Nov. 1979	Y = 103.7 + 0.784 (13)	113.9
14	Dec. 1979	Y = 103.7 + 0.784 (14)	114.7
15	Jan. 1980	Y = 103.7 + 0.784 (15)	115.5
16	Feb. 1980	Y = 103.7 + 0.784 (16)	116.2
17	Mar. 1980	Y = 103.7 + 0.784 (17)	117.0
18	Apr. 1980	Y = 103.7 + 0.784 (18)	117.8
19	May 1980	Y = 103.7 + 0.784 (19)	118.6
20	June 1980	Y = 103.7 + 0.784 (20)	119.4

Table 4.11   Deflators for Output, Material, Capital,
and Other Expenses

	Jan. 1980	Feb. 1980	Mar. 1980	Apr. 1980	May 1980	June 1980
Index	115.5	116.2	117.0	117.8	118.6	119.4
Deflator	1.000	1.006	1.013	1.020	1.027	1.034

model. In designing these forms, various factors, such as how often data are collected, the information required, computation steps, the scale of measurement, and clarity, are considered. The input and output elements must be evaluated to ensure that they can be collected periodically. The productivity analyst or coordinator must involve key personnel from the various departments in the review of the data collection system. Proposed changes to the data collection system are usually easily adopted if they have the approval of both upper management and employees. The interaction steps are shown in Figure 4.11.

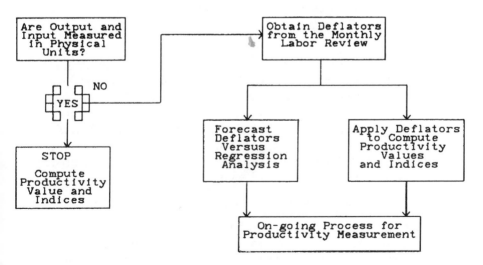

Figure 4.10   Elements involved in deflator selection.

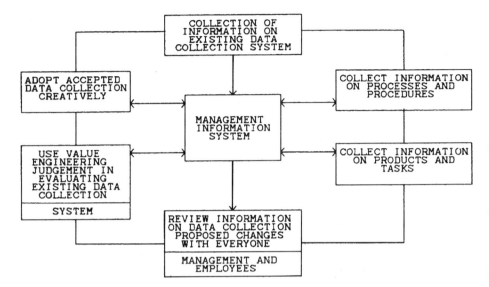

Figure 4.11   Key steps in data collection design.

Step 6: Personnel Training and Testing of Data
Collection Instruments

In this step, the productivity measurement team is trained for a reasonable period of time on how to perform the measurement of input and output elements and how to complete the productivity measurement forms, safety regulations, and operating procedures. The data collection designed in step 5 should be pretested with the involvement of both workers and managers and revised for better clarity. Usually it takes about four 1-hour sessions to acquaint the productivity team with the productivity measurement package. Figure 4.12 shows the steps needed in the training of personnel and testing of data collection systems.

Step 7: Data Collection, Synthesis, and Computations

Using the various forms designed in step 5, input and output components needed for the productivity measurement model are collected periodically for each task. The productivity analyst reviews the data collected and eliminates unnecessary information. The synthesized

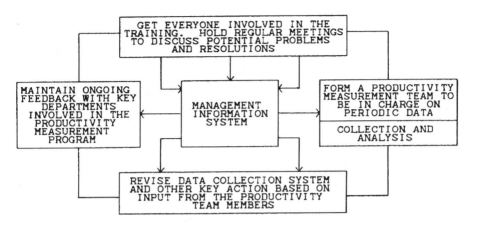

Figure 4.12 Steps in personnel training and data collection testing.

data are used to compute the values and indices for partial and total productivities. It is recommended that a data base be developed for synthesized data for the analysis of productivity trends. The example presented in the case study shows how the productivity values and indices are computed. The results may be displayed in form of bar charts or trends for easy understanding by upper management. In situations in which there are many tasks, it is recommended that productivity information be grouped into meaningful subgroups. The steps are presented in Figure 4.13.

Step 8: Trend Analysis and Interpretation of Findings

The partial and total productivities for each task obtained by period are analyzed periodically, seeking reasons for an increase or decrease. Statistical techniques, such as analysis of variance (ANOVA), can be used to interpret the data collected. By relating the partial productivities and their impact on the total productivity, specific areas that need to be investigated for improvement can be pinpointed. The productivity improvement and planning team uses information obtained for both short- and long-range improvement and planning. Key actions are developed and implemented for the problem areas. The elements involved in the trend analysis are shown in Figure 4.14.

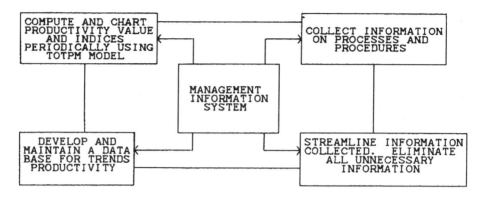

Figure 4.13   Steps in data collection synthesis and computations.

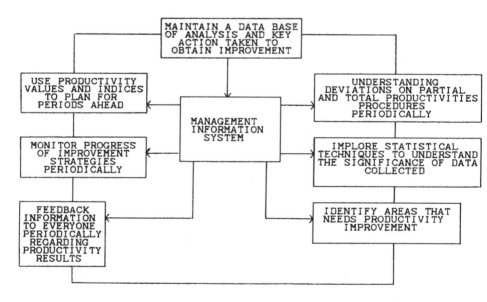

Figure 4.14   Steps in trend analysis and interpretation of findings.

4.5 CASE STUDY: COMPARISON OF THE MANUAL AND COMPUTER-AIDED METHODS OF ASSEMBLING PRINTED CIRCUIT BOARDS USING THE TASK-ORIENTED TOTAL PRODUCTIVITY MODEL*

The task-oriented productivity measurement model was used in plant Z to measure and compare partial and total productivities of the manual and computer-aided methods of assembling printed circuit boards. A formal study of the Z printed circuit board manufacturing process, procedure, cost accounting system, type of printed circuit board produced, types of operations, task performed, and the J-2000 computer operation was conducted.

The key personnel in the various departments were introduced and made to understand the purpose of the productivity measurement system. The significance of each task performed, the extent of the manual method of assembling (PCB), computer usage, and the associated input-output relationship of tasks were determined. For the purpose of our study at plant Z, a proportional contribution to the total number of printed circuit board insertions was used as an allocation criteria for overhead expenses. The allocation criteria were preferred to other allocation schemes because energy usage, machine utilization, and component preparation time varied with the number of insertions.

Plant Z manufactured various printed circuit board types that required different amounts of insertions. For example, printed circuit board Mark I type required 218 insertions as opposed to printed circuit board Mark II type with 210 insertions. Mark I therefore utilized more energy than Mark II; similarly, Mark I utilized more machine time and more labor time for the preparation of components inserted. The various printed circuit board types and the number of insertions in each printed circuit board are presented in Table 4.12. Generally, the input and output elements of the TOTPM model are deflated in order to remove the effect of price changes. If needed, the deflators are obtained from the monthly labor review published by the Bureau of Labor Statistics, Washington, D.C. The deflators are chosen based on their correlation with the various input and output elements. In this study, the deflators were not required because the study period was only

---

*Adapted from Edosomwan, 1985. Reprinted with permission.

Table 4.12 Summary of Printed Circuit Board Insertion Rate at Plant Z[a]

Printed circuit board type	Number of insertions in each printed circuit board type
Mark I	218
Mark II	120
Mark III	262
Mark IV	98

[a]MARK I was the printed circuit board type considered in our study.

5 weeks, not long enough to have significant variation in price changes.

Various forms and instruments were designed to capture the input and output elements needed for the task-oriented productivity model. In designing these forms, various factors, such as how often data were collected, the information required, computation steps, the scale of measurement, and clarity, were considered.

The subjects were trained for 1 week on how to perform both the manual and computer tasks and how to complete the productivity measurement forms. They were also taught safety regulations and plant Z operating procedures. The data collection instruments designed were pretested with five workers and two managers, and revised for better clarity. The operators were selected at random to perform the manual and computer-aided printed circuit board assembly tasks.

Using the various forms, input and output components needed for the productivity model were collected periodically, for each individual worker as well as for each group that performed the manual and computer-aided tasks. Synthesized data were used to compute the values and indices for partial and total productivities.

4.5.1 Results

The input and output components and total and partial productivities by period for both the manual and computer-aided tasks are shown in Tables 4.13 and 4.14. A summary of the research findings is also presented in Table 4.15. Examples of ANOVA tables for total and labor productivities are shown in Tables 4.16 and 4.17. At the $\alpha = 0.05$ level of

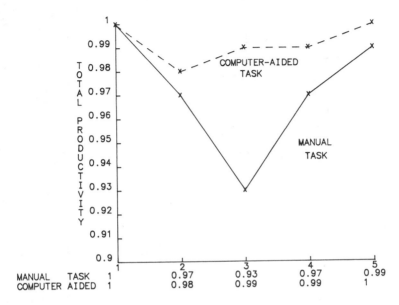

Figure 4.15 The total productivity trend for manual and computer-aided tasks.

significance, there was a significant difference in total and labor productivities between the two tasks. The computer-aided method increased total and labor productivities more than the manual method of assembling the printed circuit boards. The programming sequence of the computer-aided method enabled operators to work at a faster pace, and with the aid of the rotating bin and special handling fixtures on the machine, operators were able to assemble two printed circuit boards simultaneously. On the other hand, the manual method required operators to work at their own pace and also required more hand motion because of the way the work stations were arranged.

The computer-aided method was also found to have decreased energy, capital, computer operating expense, and other expense productivities. In addition to the energy utilized for the lighting fixtures, the computer-aided method also utilized direct electricity to run the rotary bin and the movement of the assembly fixtures. The manual method utilized energy only for the lighting fixtures. Under the lease concept approach for capital, the computer-aided method utilized more capital

Table 4.13 Output and Input Components for Manual and Computer-Aided Tasks[a]

Items	Period 1 Manual task	Period 1 Computer task	Period 2 Manual task	Period 2 Computer task	Period 3 Manual task	Period 3 Computer task	Period 4 Manual task	Period 4 Computer task	Period 5 Manual task	Period 5 Computer task
Quantity of finished PCB produced	40	55	38	53	35	54	39	53	40	54
Percentage completion of partial units of PCB	—	—	—	—	—	—	—	—	—	—
Quantity of partial PCB produced	—	—	—	—	—	—	—	—	—	—
Average selling price per unit of PCB ($)	1000	1000	1000	1000	1000	1000	1000	1000	1000	1000
Total output ($)	40,000	55,000	38,000	53,000	35,000	54,000	39,000	53,000	40,000	54,000

Input										
Labor expense input ($)	85	85	84	84	84	84	84	84	84	84
Material expense input ($)	4147	5775	3990	5565	3675	5617	4147	5513	4200	5517
Capital expense input ($)	63	73	56	67	60	71	61	70	62	73
Energy expense input ($)	86	96	86	96	86	96	86	96	86	96
Robotics expense input ($)	—	—	—	—	—	—	—	—	—	—
Computer expense input ($)	—	98	—	101	—	97	—	100	—	104
Other administrative expense input ($)	4417	5082	4417	5111	4417	5093	4417	5088	4417	5107
Total input ($)	8798	11,111	8633	10,923	8322	10,961	8795	10,851	8849	10,877

aNote: No partial units of PCB were produced during the study period. No robotic expense was incurred.

Table 4.14 Total and Partial Productivities for Manual and Computer-aided PCB Tasks

Items	Period 1		Period 2		Period 3		Period 4		Period 5	
	Manual task	Computer task	Manual task	Computer task	Manual task	Computer task	Manual task	Computer task	Manual task	Computer task
Total productivity										
Value	4.55	4.95	4.40	4.85	4.21	4.93	4.43	4.88	4.52	4.96
Index	1.00	1.00	0.97	0.98	0.93	0.99	0.97	0.99	0.99	1.00
Labor expense productivity										
Value	470.59	647.06	452.38	630.95	416.67	642.86	464.29	630.95	476.19	642.86
Index	1.00	1.00	0.96	0.98	0.89	0.99	0.99	0.98	1.01	0.99
Material expense productivity										
Value	9.65	9.52	9.52	9.52	9.52	9.61	9.40	9.61	9.52	9.79
Index	1.00	1.00	0.97	1.00	0.97	1.01	0.97	1.01	0.97	1.03
Capital expense productivity										
Value	634.92	753.42	678.57	791.04	583.33	760.56	639.34	757.14	645.16	739.73
Index	1.00	1.00	1.07	1.05	0.919	1.01	1.01	1.00	1.02	0.98

Energy expense productivity										
Value	465.11	572.92	441.86	552.08	406.98	562.50	453.48	552.08	465.12	562.50
Index	1.00	1.00	0.95	0.96	0.88	0.98	0.97	0.96	1.00	0.98
Robotics expense productivity										
Value	—	—	—	—	—	—	—	—	—	—
Index	—	—	—	—	—	—	—	—	—	—
Computer expense productivity										
Value	—	561.22	—	524.75	—	556.70	—	530.00	—	519.23
Index	—	1.00	—	0.94	—	0.99	—	0.94	—	0.93
Other administrative expense productivity										
Value	9.06	10.82	8.60	10.37	7.92	10.60	8.83	10.42	9.06	10.57
Index	1.00	1.00	0.95	0.96	0.87	0.98	0.97	0.96	1.00	0.98

than the manual method to support tools, maintenance, programming, safety, and indirect effort supervision costs. This impacted both computer operating expenses and other expense productivities. No significant difference in material productivity was observed. Although the results have been used to compare the computer-aided and manual methods of assembling printed circuit boards, the trends for each task by period are also essential, especially for investigating areas that need productivity improvement. The total productivity trend for the manual and computer-aided tasks is presented in Figure 4.15. In the situation of interest, the decision maker is faced with using the computer-aided method to accept gains in total and labor productivities but investigating and continuing to improve the other partial productivities (energy, capital, computer operating expense, and other expenses) or using the manual method to assemble the printed circuit boards in an environment in which labor is inexpensive.

4.6 CASE STUDY: ROBOTIC DEVICE IMPACT ON TOTAL PRODUCTIVITY IN PRINTED CIRCUIT BOARD ASSEMBLY TASK*

4.6.1 Research Site and Methodology

Data were collected at a manufacturing plant that introduced and used robotic device J-1515 in the assembling of printed circuit board. Manual insertion of the electronic components was considered difficult and time consuming. Comparison of the total productivity of both the manual and robotic device method was needed to select the best production method.

4.6.2 Description of the Manual Method of Assembling Printed Circuit Boards

The manual method of assembling printed circuit boards was a process in which the production worker manually inserted components, such as resistors, diodes, modules, transistors, and capacitors, into an empty circuit board or raw card.

Source: From Edosomwan, J. A., 1986. A methodology for assessing the impact of robotics on total productivity in an assembly task. Annual International, Industrial Engineering Conference Proceedings, Dallas, Texas, May 9–11. Reprinted with permission.

Table 4.15 Summary of Research Findings for Case Study[a]

Variable	Difference in mean productivity values between manual and computer task	Impact on total and partial productivities		Significant at = 0.05
Total productivity	0.49	C+	M−	Yes
Labor expense productivity	182.92	C+	M−	Yes
Material expense productivity	0.09	C*	M*	No
Capital expense productivity	124.12	C−	M+	Yes
Energy expense productivity	113.91	C−	M+	Yes
Computer operating expense productivity	538.38	C−	M+	Yes
Other administrative expense productivity	1.87	C−	M+	Yes

[a]C+ = computer-aided task had positive impact on productivity. C− = computer-aided task had negative impact on productivity. M+ = manual task had positive impact on productivity. M− = manual task had negative impact on productivity. * = no significant difference in productivity was observed.

The work area was in an approved electrostatic discharge area that consisted of a grounded work bench, chair, tools, and a light box. The manual assembly process started with the verification of paperwork and parts. The component placement list (CPL), picklist, routing, part numbers, engineering change level, serial numbers, templates, and any special instructions were checked before assembly. The foil, which was a blueprint of component insertions, was placed on the lightbox. The foil contained the outline for two boards. The components were manually inserted, one at a time, according to the foil. Upon completion of both cards, they were each verified with the CPL and template. If any rework or scrap was to be done it was performed in a manner similar to that of insertion. When the printed circuit boards were completely assembled and verified, they were manually placed into trays.

Table 4.16   Total Productivity—Manual Task Versus Computer-Aided Task (ANOVA) (r = 0.05)[a]

Source of variation (SV)	Degrees of freedom (DF)	Sum of squares (SS)	Mean squares (MS)	Observed F values	Critical °F values
Between task	1	1.60	1.60	160.0	7.71
Between days	4	0.05	0.01	1.0	7.71
Error	4	0.03	0.01	—	—
Total	9	1.68	—	—	—

[a]Between task conclusion: Since F data = 160 > °F critical = 7.71. We reject H_o; and conclude that there is significant difference in total productivity between manual task and computer task. Computer task increases total productivity more than manual task. Between days conclusion: Since F data = 1.0 > °F critical = 7.71. We accept H_o and conclude that there is no significant day effect for manual task and computer task.

4.6.3   Description of the Robotic Device Method of Assembling Printed Circuit Boards

The robotic device method of printed circuit board assembly was a process in which the production worker assembled printed circuits using the robotic device J-1515 and controller. The robot was used to insert rectangular chip packages varying from ½ inch to 1½ inches onto a partially populated printed circuit board.

The work area was in an approved electrostatic discharge area that consisted of robotic devices J-1515 and controller, module insertion heads for ½ inch and 1½ inch modules, machine base, clinch unit, application control, cabinet, industrial program computer, module feeder, manual card feeder, card carriers, and a set of inspection templates.

The robotics method of assembly begins in a manner similar to the manual method. The process begins with verification of the job paperwork and parts. The paperwork is the same as that used in the manual process, with the exception of any special instructions unique to the operation of the robot.

Once the job paperwork and parts are verified, the operator is ready to begin the setup of the tool, in the following steps. Ensure that all

Table 4.17   Labor Productivity—Manual Task Versus Computer-Aided Task (ANOVA) (r = 0.05)[a]

Source of variation (SV)	Degrees of freedom (DF)	Sum of squares (SS)	Mean squares (MS)	Observed F values	Critical F values
Between task	1	18.28	18.28	261.1	7.71
Between days	4	0.34	0.08	1.14	7.71
Error	4	0.28	0.07	—	—
Total	9	18.90	—	—	—

[a]Between task conclusion: Since F data = 261.1 > °F critical = 7.71. We reject H_o and conclude that there is significant difference in labor productivity between computer task and manual task. Between days conclusion: Since F data = 1.14 > °F critical = 7.71. We accept H_o and conclude that there is no significant day effect in labor productivity from computer task and manual task.

switches are in the on position, as outlined in the operating procedure; obtain the proper workboard holder, and load onto the fixture. Once the fixtures are loaded, the operator is ready to load the PCB assembly program into memory in preparation for assembly. The program contains the insertion pattern and indicates which part number is to be loaded in the proper input channel. The program would then position the X–Y table into proper position and guide the robot arm to the proper channel to pick up a part and insert the part into the PCB.

4.6.4   Measurement Method for Productivity

The task-oriented total productivity measurement model presented in Section 4.4 was used in this study. The implementation methodology described in Section 4.4 was followed.

4.6.5   Results

As shown in Figures 4.16 through 4.20, when the manual method of assembling printed circuit boards was changed to the robotic device method there was an increase in labor and total productivity. The robotic device method was able to work consistently faster than the manual method. The programming sequence of the robotic device eliminated extra motions that were performed using the manual method. Excessive transportation, handling, grasping, and reaching motions were eliminated using the robotic device method. Although the robotic device used a consistent pace, the manual method required operators to work at their own pace and also required more motions because of the way the work envelope was designed. The robotic device method was found to have decreased energy, capital, robot operating expense, data processing, and other productivity expenses. More electricity was utilized to support the mechanical motion of the robot movements. The manual method utilized electricity only for the lightbox. The robotic device required more indirect effort in supervision and more maintenance, safety procedures, programming support, and tooling costs. This accounted for the decrease in most of the partial productivities. However, the gains in labor productivity were large enough to offset the impact of the operating costs. It is important to improve the design of the robot to minimize the operating cost; if not, in the long run total productivity may be significantly affected.

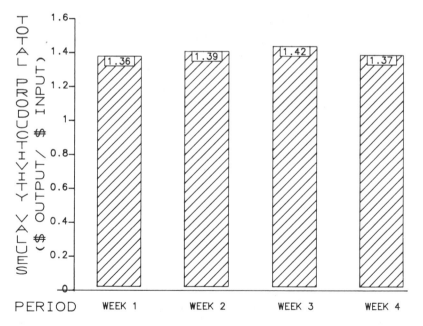

Figure 4.16   Total productivity values for the manual PCB task.

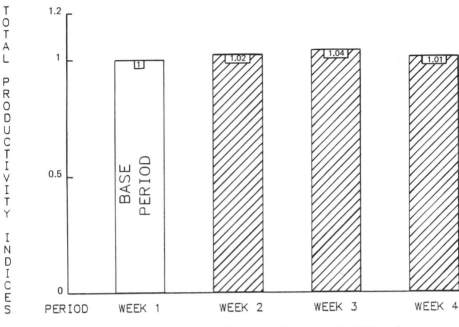

Figure 4.17 Total productivity indices for the manual PCB task.

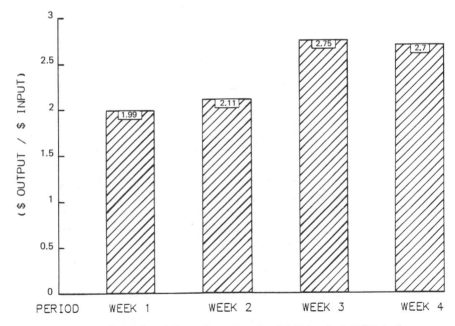

Figure 4.18 Total productivity values for the J-1515 robot PCB task.

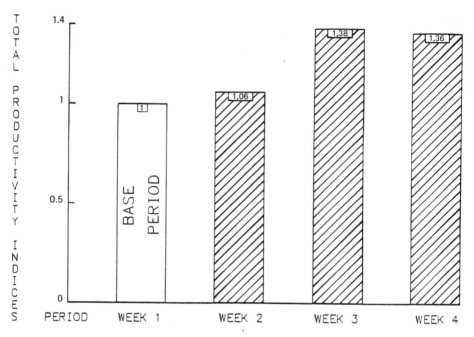

Figure 4.19   Total productivity indices for the J-1515 robot PCB task.

This is perhaps the first attempt to assess the impact of a robotic device on total productivity. In the past, labor productivity and the rate of return on investment (ROI) has been used to justify the application of technology, such as robotic devices. These measures are partial measures that do not consider the total cost of doing business. Variable operating cost, has often been neglected. The total productivity measurement approach presented here has a wide range of application in assessing the productivities of other types of technology, such as computer-aided manufacturing. Much work is needed to design and produce improved robotic devices that have perfect error recovery mechanisms, less costly safety procedures, fewer programming steps, and less indirect effort support. A close link between the systems designer and manufacturing personnel through early manufacturing involvement (EMI) in the design phase will also help minimize system design problems.

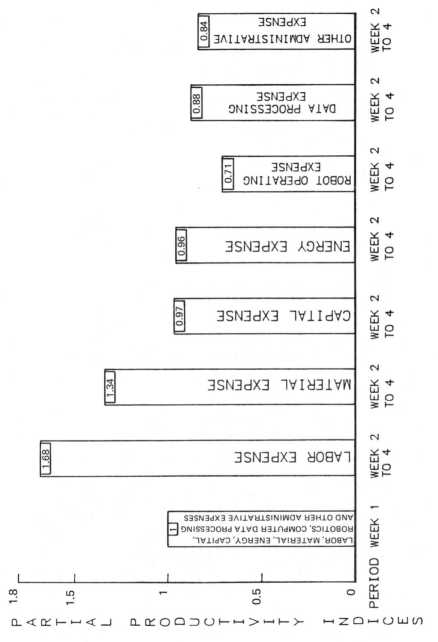

Figure 4.20 Partial productivity indices for the J-1515 robot PCB task.

4.7 CASE STUDY: COMPUTER-AIDED MANUFACTURING IMPACT ON TOTAL PRODUCTIVITY IN PCB ASSEMBLY TASK

4.7.1 Research Site and Methodology

Data were collected at a manufacturing plant that introduced and used specific types of computers (J-2000 and J-2000-1) to assemble printed circuit boards. The computer-aided printed circuit board assembly task was concerned with the use of a logistical application program to assist production workers in the insertion of small electronic components, such as transistors, diodes, resistors, and capacitors, into a blank printed circuit board. The need for the computer-aided method arose from the fact that manual insertion of the electronic components was difficult and time consuming. Both the manual and computer-aided methods of assembling printed circuit boards were performed in an electrostatic discharge area.

4.7.2 The Subjects and the Research Sequence

Two difference groups of five subjects each were studied. The first consisted of five normal subjects (two males and three females) with a mean age of 32 years (range, 27–40). The second group consisted of five normal subjects (two males and three females with a mean age of 35 years; range, 26–43). All subjects were selected after passing a physical examination and after a careful review of their response to the health screening. In order to ascertain that the research results were not biased by the effect of the order in which tasks were performed, the research was designed to test for the effect of the order of performance. In the two case studies of interest, the participants were divided into two groups. In the first case study, group 1 performed the manual task initially, followed by the computer task. Group 2 initially performed the computer task followed by the manual task. The subjects in group 1 were assigned code letters A, B, C, D, and E. Similarly the subjects in group 2 were assigned code letters F, G, H, I, and J. In the second case study, group 1 performed the computer task using J-2000 initially, followed by the computer task using J-2000-1. Group 2 performed the computer task using J-2000-1 initially, followed by the computer task using J-2000. Using a random selection sequence, the participants were trained to assemble the printed circuit boards manually and by computer. Following this, data on partial and total productivities were collected during an actual production period of ten weeks. The research sequence is

Table 4.18   Research Sequence

Case study number	Group	Task sequence	Group population
Case study 1 research sequence	Group 1	Manual task, computer task (J-2000)	5
	Group 2	Computer task (J-2000), manual task	5
		Total	10
Case study 2 research sequence	Group 1	Computer task (J-2000), computer task (J-2000-1)[a]	5
	Group 2	Computer task (J-2000-1), computer task (J-2000)	5˙
		Total	10

[a]The J-2000-1 is a modified version of the J-2000 computer.

presented in Table 4.18. The task-oriented total productivity measurement model and the implementation methodology describe in Section 4.4 was applied.

4.7.3   Results

The results showed that when the J-2000 system was changed to the J-2000-1 computer system there were increases in total productivity and labor, materials, and energy productivities. The productivities related to capital, computer operating expense, data processing expense, and other administrative expenses decreased. The results of the case study are shown in Figures 4.21 through 4.24.

4.7.4   Computer Systems Problems Encountered During Implementation

In both computer systems J-2000 and J-2000-1, the computer speed was fixed. The same Δt was used for types of components. Δt was the time interval allowed for operators to insert a component into the printed

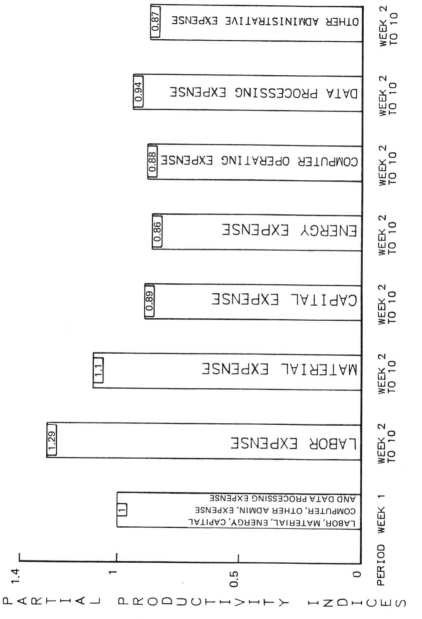

Figure 4.21 Partial productivity indices for the J-2000 computer-aided task.

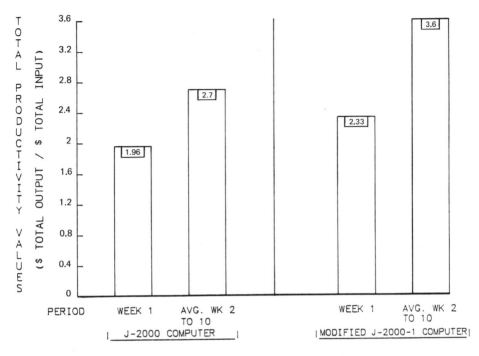

Figure 4.22   Total productivity values for the J-2000 and J-2000-1 computer-aided tasks.

circuit board. This created a control problem for operators, especially in situations in which the component lead length and sizes were different. The logistical application programs of the J-2000 and J-2000-1 computer systems also malfunctioned. The common errors observed were (1) feeding an incorrect part to the operator, (2) providing a wrong assembly sequence, (3) wrong coordinates leading to parts jamming the fixtures, (4) providing the wrong correction code to the operator, (5) shining a light beam locator at an invalid coordinate, (6) mechanical breakdown of the system, and (7) operator-induced error causing the system to be down. The workers perceived that these systems-related problems created significant control and interdependence problems that affected their job satisfaction and caused psychological stress. System design problems were resolved through three main strategies: (1) involvement of users and their recommendations in the design stage of

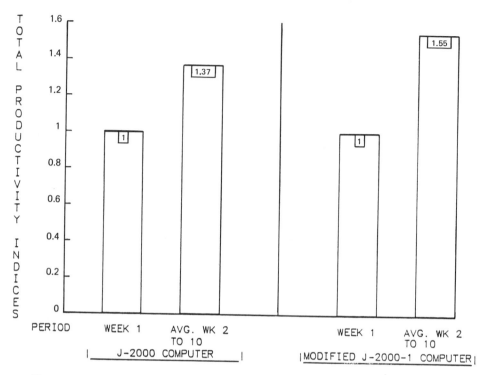

Figure 4.23　Total productivity indices for the J-2000 and J-2000-1 computer-aided tasks.

the system; (2) modification of the computer-based systems parameters to allow control for workers, for example, a variable Δt was provided in the modified computer system after this investigation recommendation; and (3) computer system commands, variables, and parameters were enhanced to enable the system to be user friendly.

4.7.5　Problems Encountered During Implementation

The problems presented here are mainly intended to form a guide to companies setting up a productivity measurement program for the first

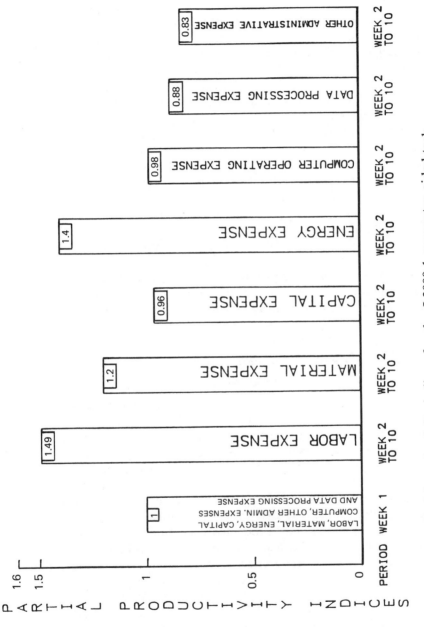

Figure 4.24 Partial productivity indices for the J-2000-1 computer-aided task.

time. It should be noted, however, that the problems associated with starting a productivity measurement program may vary depending on the type of production environment. Based on the implementation of the task-oriented total productivity model at various plants, we found that the existing data collection system could not provide the input and output components needed for the TOTP model. A computer program was developed to enable us to collect input and output by task at the time of production. A session was held with the cost accounting department; this resulted in a slight modification of the cost accounting system. As with any other such study, employees have the initial impression that a study brings layoffs, change in procedures, and other problems. It was not unusual for us to notice similar feelings and a resistance to change at the beginning of our study, but these faded eventually as the objectives of our study were communicated to and supported by both management and employees. Initially, data collection was very difficult because we had to rely on specific persons who had to add the task of helping in data collection to their normal duties. The team work approach adopted to implement the model eliminated the "it's not my problem" syndrome. Specific tasks were assigned to specific people, and meetings were held regularly to discuss and resolve potential problems.

4.7.6   Productivity Measurement Project Maintenance

Validation of a total productivity measurement model takes into account all the measurable factors of input and output. Although the model was found to be sensitive in the various test cases of the printed circuit board assembly tasks, a potential exists for the application of the model in other types of tasks performed in both manufacturing and service organizations. However, minor modifications will have to be made in model notations and data collection instruments. A computer program is also available to ease the computational burden.

In order to maintain the productivity measurement program once it has already been established using the task-oriented total productivity model, the following are needed: (1) data should be collected periodically; this will depend upon whether the productivity indices are established on a monthly, quarterly, or yearly basis; (2) productivity indices are computed for each task and for the firm and analysis of the trends for specific areas to be investigated for improvement; (3) tasks are compared in the same base period; (4) development of historical data on new tasks to permit sales, costs, and profit analysis before the tasks are introduced to the productivity measurement program; (5) productivity trend patterns should be compared with the actual profit figures by period; and (6) training sessions should be conducted with the key personnel in the various departments that will be involved in the maintenance of the productivity measurement program.

4.7.7 Conclusions

This is perhaps the first research that has investigated the impact of computer-aided manufacturing on total productivity. More work needs to be done in understanding the full implications of computer-aided manufacturing on productivity other than the one environment studied. The task-oriented total productivity measurement utilized in the case studies can also be applied in other types of tasks with minor modifications in the data collection instruments. Based on the findings of this research, system designers and managers of automated tasks are provided with the following recommendations. (1) Involve all potential users of computer-based systems during the design phase. (2) Provide flexibility on computer-based systems that allows a stress-free human/computer interface. (3) Provide the ability to advise, alert, or warn users of potential events on the computer-based systems. (4) Evaluate the system performance to decrease costly additional safety and operational procedures. (5) Evaluate the system performance for cost minimization in energy, capital, data processing, and indirect effort supervision costs.

4.8 TECHNOLOGY-ORIENTED TOTAL PRODUCTIVITY MEASUREMENT SYSTEM

In this era of technological explosion, companies are faced with the challenge of how to maintain technological leadership and a competitive edge, increase revenue, diversify products, keep abreast of the market place, and provide the best possible technology for the customer. However, the life cycle of the new technologies has continued to be shortened from about a 10-year span in 1960 to about a 2-year span in 1986. How to respond to the short life cycle of technology and manage it for productive growth is a significant issue. Edosomwan (1986) pointed out that one way to do this is through a careful understanding and management of the input-output relation in the stages of the technology life cycle. Following this, developing and implementing a technology-oriented productivity measurement system can have the following benefits.

1. A technology-oriented productivity measurement system improves resources planning for all phases of the technology life cycle. It provides common measures and concepts to compare similar or different types of technology.
2. Technology-oriented productivity measurement systems create the awareness required for effective supervision of all actions to be taken and improves decision-making through better understanding of the actions already taken to address a given problem in each of the phases of the technology life cycle.
3. Technology-oriented productivity measurement systems provide a basis for the firm to make decisions on when to phase out a given technology. The productivity values and indices provide a way of detecting deviations from established standards on a timely basis so that something is done about such deviations. In situations in which the performance of the work groups, technology, task, customer, departments, projects, and the firm as whole is to be measured, the technology-oriented measurement system can be implemented with minor modifications for the definition of inputs and outputs and allocation criteria for overhead expenses. The major significant difference between previous productivity models and the technology-oriented total productivity measurement model is that it recognizes a new technological version in which development expenses, retraining expenses for new technologies, computer operating expenses, and robotic operating expenses are important additions to the input components. New ideas and other associated values have also been considered output components. Also, productivity is measured in all phases of the technology for the first time. The measures derived from this model are in the form of an index obtained by considering all phases of the technology life cycle shown in Table 4.19. The input and output components of the TOTPM model are shown in Figures 4.25 and 4.26. The key definitions associated with the TOTPM model are as follows:

Table 4.19 Technology Life Cycle Phases[a]

Phase description	1. Project proposal study and definition of ideas	2. Design	3. Development	4. Manufacturing verification, production, and sales
End objective at each phase	Determine feasibility, requirements, initial specifications cost, quantities, and estimated revenue and profits	Refine specifications, build models, define market support and verify costs revenue and profits, commit ideas and model to marketing	Finalize specifications costs, test manufacturability, release for announcement, and commit price and revenue	Verify manufacturing capability; produce, deliver, and install; continue to forecast future demand; produce until loss of demand indicates end of life
Significant input components	Labor expenses Other administrative expenses	Development expenses Labor, materials Capital Energy Other administrative expenses	Development expenses Labor, capital, material Energy Other administrative expenses	Labor, material Capital, energy Computer operating expense Robotic operating expense Retraining expenses Other administrative expenses
Significant output components	New ideas in monetary terms ($ value)	Finished model Partial model in monetary terms ($ value)	Finished unit prototype Partial unit prototype in monetary terms ($ value)	Partial units produced Finished units produced Other output associated with units produced in monetary terms ($ value)

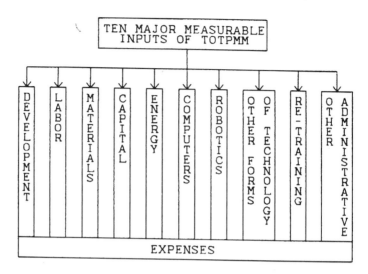

Figure 4.25 Input components considered in the technology-oriented total productivity measurement model.

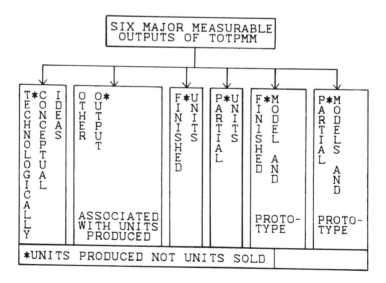

Figure 4.26 Output components considered in the technology-oriented total productivity measurement model.

4.8.1   Notations for the Derivation of Productivity Values and Indexes

Let

i = technology type (i = 1, 2, 3, ... m)

j = technology life cycle phase (j = 1, 2, 3, 4)

k = technology development-manufacturing-service site (k = 1, 2, 3, ... n)

t = study period (t = 1, 2, 3, ... p)

TCI_{ijkt} = total value of new ideas of type i, produced in phase j, in site k, in period t

FMP_{ijkt} = total value of finished model and prototype of type i, produced in phase j, in site k, in period t

PMP_{ijkt} = total value of partially completed model and prototype of type i, produced in phase j, in site k, in period t

FUP_{ijkt} = total quantity of finished units of technology of type i, produced in phase j, in site k, in period t

PUP_{ijkt} = total quantity of partial units of technology of type i, produced in phase j, in site k, in period t

PCT_{ijkt} = percentage completion of partial units of technology of type i, produced in phase j, in site k, in period t

SPT_{ijkt} = base period selling price per unit for a unit of technology of type i, produced in phase j, in site k, in period t

TOT_{ijkt} = total quantity of output of technology of type i, produced in phase j, in site k, in period t

DET_{ijkt} = development expense input in monetary terms utilized to produced technology of type i, in phase j, in site k, in period t

LRT_{ijkt} = labor hours utilized to produce technology of type i, in phase j, in site k, in period t

LRR_{ijkt} = labor rate per hour utilized to produce technology of type i, in phase j, in site k, in period t

MET_{ijkt} = materials and purchase part expense utilized to produce technology of type i, in phase j, in site k, in period t

CET_{ijkt} = capital-related expense (includes fixed and working capital, such as cash, accounts receivable, tools, plant, and buildings) utilized to produce technology of type i, in phase j, in site k, in period t (capital is computed using lease value concept)

EET_{ijkt} = energy-related expense (includes electricity, solar energy, water, coal, and gas); input utilized to produce technology of type i, in phase j, in site k, in period t

CRE_{ijkt} = variable computer-related expense input utilized to produce technology of type i, in phase j, in site k, in period t

RRE_{ijkt} = variable robotics-related expense input utilized to produce technology of type i, in phase j, in site k, in period t

OFT_{ijkt} = variable expense from other forms of technology utilized to produced technology of type i, in phase j, in site k, in period t

RET_{ijkt} = retraining expense utilized for technology of type i, in phase j, in site k, in period t

OET_{ijkt} = other administrative expense utilized to produced technology of type i, in phase j, in site k, in period t

TBT_S = base period time, the reference period to which output and input in monetary terms is reduced; thus, the total productivity of a task expressed as dollar output per dollar input in constant dollars

TPT_{ijkt} = total productivity of technology of type i, in phase j, in site k, in period t

TFT_{ijkt} = total factor productivity of technology of type i, in phase j, in site k, in period t

PPT_{ijkt} = partial productivity of technology of type i, in phase j, in site k, in period t

OOT_{ijkt} = other output associated with units produced of technology of type i, in phase j, in site k, in period t

TIT_{ijkt} = total input utilized to produced technology of type i, in phase j, in site k, in period t

The total and partial productivities are expressed as follows:

$$TPT_{ijkt} = \frac{\text{Total measurable output produced of technology of type i, in phase j, in site k, in period t.}}{\text{Total measurable input utilized to produce technology of type i, in phase j, in site k, in period t}} \qquad (4.6)$$

$$TPT_{ijkt} = \frac{TOT_{ijkt}}{TIT_{ijkt}} \tag{4.7}$$

$$TPT_{ijkt} = \frac{\begin{array}{c} FMP_{ijkt} + TCI_{ijkt} + PMP_{ijkt} + FUP_{ijkt}SPT_{ijkt} \\ + PUP_{ijkt}PCT_{ijkt}\,SPT_{ijkt} + OOT_{ijkt} \end{array}}{\begin{array}{c} DET_{ijkt} + LRT_{ijkt}LRR_{ijkt} + MET_{ijkt} + CET_{ijkt} \\ + EET_{ijkt} + CRE_{ijkt} + RRE_{ijkt} + OFT_{ijkt} + RET_{ijkt} \\ + OET_{ijkt} \end{array}} \tag{4.8}$$

Partial productivity considers only one input factor. Total factor productivity considers only labor and a capital input factor. For example, partial productivity with respect to labor input is expressed as

$$PPT_{ijkt} = \frac{\begin{array}{c} TCI_{ijkt} + FMP_{ijkt} + PMP_{ijkt} + FUP_{ijkt}SPT_{ijkt} \\ + PUP_{ijkt}PCT_{ijkt}SPT_{ijkt} + OOT_{ijkt} \end{array}}{LRT_{ijkt}LRR_{ijkt}} \tag{4.9}$$

where LRT_{ijkt} and LRR_{ijkt} are labor terms.

4.8.2   Allocation Criteria for Overhead Expenses

The following allocation criteria are suggested.

1. Complexity factor criteria. The complexity of the technology, product, or task is used to determine a proportional contribution to the total quantity produced. Complexity in this case means such items as the number of assemblies within a given technology, the insertion rate, and the kitting time.
2. Quantity produced proportionality criteria. The proportional contribution to total quantity produced is used to allocate overhead expenses. This is applicable in situations in which there is more than one type of technology, product, or task.
3. Direct hours allocation criteria. This approach requires daily direct labor recording and the derivation of burden rate from net expense and direct hours. Therefore, the cost of a product includes labor and burden. Allocation of overhead is done using direct hours and one burden rate.

4.8.3   Implementation Methodology

The implementation steps for the technology-oriented total productivity measurement (TOTPM) model are briefly described in Figure 4.27. A teamwork approach among departments, managers, and employees is recommended to facilitate better productivity measurement.

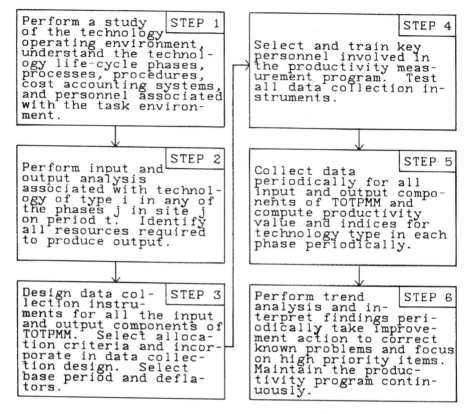

Figure 4.27   Implementation steps for the TOTPM model.

The implementation steps presented in Figure 4.27 are based on the two case studies presented in this Chapter. Additional work may be required depending on the type of technology, organizational environment, and data collection system already available.

4.8.4   Case Studies: TOTPM in Two Technology Producing Companies

The TOTPM model was used to measure and compare the productivities of types of technology in an organization that produces technology to customer order. The first case study involved a situation of a computer technology monitored throughout its life cycle. The second case study involved a situation of a robotic device technology monitored throughout its life cycle. In both case studies, two phases of each technology type were analyzed and studied. Historical data on input and output were used, especially in the development and project proposal phases. The implementation methodology specified in Figure 4.27 was followed.

Results and Discussion

As shown in Figures 4.28 and 4.29, the TOTPM model enabled us to compare the total productivity indices of two types of computers and robotic devices in different stages of their life cycles. By examining the total productivity trend of the model C212JX computer over its life cycle, improvement actions were put in place to improve the perfor-

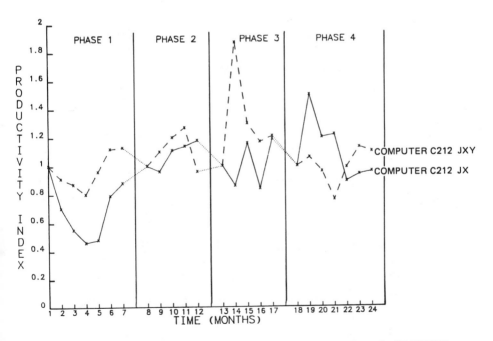

Figure 4.28   Total productivity indexes for C212JX and C212JXY computers.

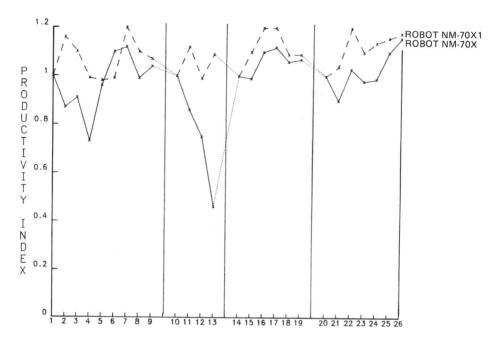

Figure 4.29   Total productivity indexes for robots NM-70X and NM-70Y.

mance trend in model C212JXY. Similar actions were taken for the robotic device models NM-70X and NM-70X. In both technologies, the series of engineering changes (EC) and specification revisions had a significant negative impact on productivity. Phase I of C212JX and C212JXY was significantly affected by labor productivity. In both technologies, energy expense, computer operating expenses, robotic operating expense, and other administrative expenses influenced the total productivity in all phases.

Problems Encountered During Implementation

Data collection and tracking was the greatest problem encountered during implementation. Because of the various phases involved in the technology life cycle, a large data base had to be set up to collect all the input and output components needed for the TOTPM model. A computer program was written to compute the values needed for the computation of productivity. In the proposal design and development phases, it was somewhat difficult to identify the contribution of a given input to the output obtained. This problem was resolved through an appropriate tracking mechanism implemented to record input and output elements. Resistance to the productivity measurement program was overcome through a teamwork approach. Sessions were held to explain the purpose and benefits of the study to everyone. It must be pointed out that the problems encountered during implementation may vary from one organization to another depending on technology types, the organizational climate, and other factors.

Conclusion

This may be the first attempt to develop and implement a technology-oriented total productivity measurement model that recognizes all phases of technology life cycle stages and total measurable input and output components. The model has also been proven to be sensitive in the two cases presented. For existing companies trying to implement this type of measurement system, it is suggested that they begin with detailed analyses of technology input and output and historical data for at least 2 years. New companies have the advantage of proceeding immediately to appropriate tracking of the technology input and output elements. The TOTPM model is a powerful model with strong potential for application to various product groups, tasks, and projects in both manufacturing and service organizations. For each potential implementation case, minor modification may have to be made in data collection instruments. The TOTPM model can also be used for product and technology equipment selection. Application of the model calls for ongoing data collection, computation, comparison monitoring, and analysis of productivity trend patterns. Only in such situations do firms realize the optimum benefits for the use of the model in a productivity measurement program.

4.9 THE COMPREHENSIVE PRODUCTIVITY MEASUREMENT MODEL

Both the task-oriented and technology-oriented total productivity measurement models can be extended to a comprehensive productivity measurement (CPM) model. The CPM model will have to take into account subjective factors, such as the impact of motivation on productivity and other intangibles. The CPM model is only recommended in situations in which there are meaningful approaches for quantifying the intangibles of input and output components. Weighting schemes for factors that affect input and output and managerial judgment are often used to quantify intangibles.

The next chapter discusses how statistical process control can be a valuable tool for improving productivity and quality.

REFERENCES

Aggrawal, S. C., 1979. A study of productivity measures for improving benefit-cost ratios of operating organizations, *Proceedings of the Fifth International Conference on Production Research*, Amsterdam, The Netherlands, August 12–16, pp. 64–70.

American Productivity Center, 1978. Productivity and the Industrial Engineer. Region VIII, AIIE Conference, Chicago, October 27 (seminar notes).

Craig, C. E. and C. R. Harris, 1972. *Productivity Concepts and Measurement—A Management Viewpoint*. Unpublished Master's Thesis, M.I.T., Cambridge, Massachusetts.

Craig, C. E., and C. R., Harris, 1973. Total productivity measurement at the firm level. *Sloan Management Review*, vol. 14, no. 3, pp. 13–29.

DeWitt, F., 1970. Technique for Measuring Management Productivity, *Management Review*, vol. 59, pp. 2–11.

DeWitt, F., 1976. Productivity and the industrial engineer. *Industrial Engineering*, vol. 8, no. 1, pp. 20–27.

Dhrymes, P. J., 1963. Comparison of Productivity behavior in manufacturing and service industries. *Rev. Econ. Stat.*, vol. 45, no. 1, pp. 64–69.

Edosomwan, J. A., 1980. Implementation of a total productivity model in a manufacturing company. Master's Thesis, Department of Industrial Engineering, University of Miami, July.

Edosomwan, J. A., 1985a. A methodology for assessing the impact of computer technology on productivity, production quality, job satisfaction, and psychological stress in a specific assembly task. Doctoral dissertation, Department of Engineering Administration, The George Washington University, Washington, D.C. 20052, January.

Edosomwan, J. A., 1985b. A task-oriented total productivity measurement model. Proceedings for First International Electronic Assembly Conference, Santa Clara, California, October 7–9.

Edosomwan, J. A., 1985c. A method for assessing the impact of computer-aided manufacturing on productivity, job satisfaction, and psychological stress in an assembly task. IIE Conference, December.

Edosomwan, J. A., 1986. The impact of computer-aided manufacturing on total productivity. 8th Annual Conference: Computers and *Industrial Engineering Journal*, March 20–22, Orlando, Florida.

Edosomwan, J. A., and D. J. Sumanth, 1985. *A Practical Guide for Productivity Measurement in Organizations: Working Manual.*

Edosomwan, J. A. 1986. "A Technology-oriented total productivity measurement model" working paper for the first International Conference on productivity research, Miami, Florida. February 5–7, 1987.

Gold, B., 1976. Tracing gaps between expectations and results of technological innovation: The case of iron and steel. *Journal of Industrial Economics*, September,

Hines, W. W., 1976. Guidelines for implementing productivity measurement. *Industrial Engineering*, vol 8, no. 6, pp. 40–43.

Kendrick, J. W., 1984. *Improving Company Productivity: Handbook with Case Studies.* The John Hopkins University Press, Baltimore.

Kendrick, J. W., and D. Creamer, 1965. *Measuring Company Productivity: Handbook with Case Studies.* Studies in Business Economics, No. 89, National Industrial Conference Board, New York.

Mali, P., 1978. *Improving Total Productivity: MBO Strategies for Business, Government, and Non-Profit Organizations.* John Wiley and Sons, New York.

Mao, J. C. T., 1965. Measuring productivity of public urban renewal expenditures. *Michigan Business Review*, vol. 17, pp. 30–34.

Melman, S., 1956. *Dynamic Factors in Industrial Productivity.* John Wiley and Sons, New York.

Mundel, M. E., 1976. Measures of productivity. *Industrial Engineering*, vol. 8, no. 5, pp. 32–36.

Siegel, I. H., 1976. Measurement of company productivity. In *Improving Productivity Through Industry and Company Management.* National Center for Productivity and Quality of Working Life, Washington, D.C., Series 2, pp. 15–25.

Stewart, W. T., 1978. A yardstick for measuring productivity. *Industrial Engineering*, vol. 18, no. 2, pp. 34–37.

Sumanth, D. J., 1979. Productivity measurement and evaluation models for manufacturing companies. Doctoral dissertation, Department of Industrial Engineering, IIT, Chicago, August.

Taylor, B. W., III, and R. K. Davis, 1977. Corporate productivity—getting it all together. *Industrial Engineering*, vol. 9, no. 3, pp. 32–36.

Tucker, S. A., 1961. *Successful Management Control by Ratio Analysis*, McGraw-Hill, New York.

Turner, J. A., 1980. Computers in bank clerical functions: Implication for productivity and the quality of working life. Doctoral dissertation, Columbia University, New York.

U.S. Department of Labor, Bureau of Labor Statistics, 1980. *Monthly Labor Review*, January, pp. 40–43.

5

Productivity and Quality Improvement Through Statistical Process Control

This chapter presents the concepts and techniques of statistical process control (SPC) and its usefulness in improving productivity and quality in the business environment. A step-by-step methodology for implementing SPC and the design of the experiment are presented. A case study that shows the application of the SPC technique in a group technology production environment is presented. A procedure for designing an experiment is also offered. Statistical tools for data analysis are described.

5.1 PROCESS CONTROL DEFINITION

Process control is a state whereby statistical inference techniques are used to monitor and control a specified process in order to achieve improved quality and gains in productivity. The control concept utilizes both historical and present technical knowledge of the process in understanding cause-and-effect relationships combined with statistical techniques to control and minimize defects. The implementation of process control concepts and techniques is achieved by providing a control system for defect and error detection, a control system for defect and error analysis, and a control system for defect and error correction.

5.2   A PROCESS CONTROL SYSTEM DEFINITION

A process control system is a feedback mechanism that provides information about the process characteristics and variables, process performance, action on the process inputs, transformation process, and action on the output. The major components of a process control system are presented in Figure 5.1.

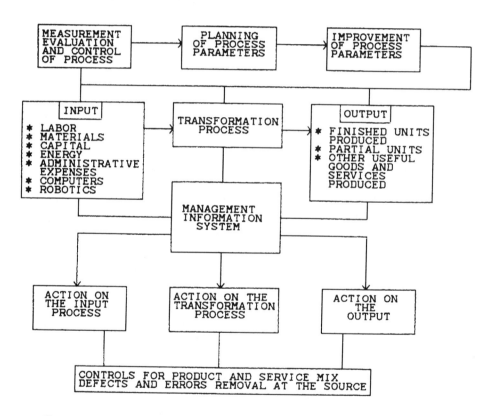

Figure 5.1   Components of a process control system.

5.3 THE KEY REQUIREMENTS OF PROCESS MANAGEMENT

In order to improve the quality of products or services from a process, the following basic requirements are necessary.

1. The management philosophy must be one that is geared toward continuous quality improvement at the source of production or service. Process control is not a one-time action. It is a continuous feedback mechanism that involves both management and employees in producing good products and services.
2. There must be a system for measurement, control, evaluation, planning, and improvement.
3. There must be a data base for the storage, retrieval, and access of information about the process.
4. There must be a clear definition of process parameters and variables. Input and output elements must be clearly identified.
5. All process parameters must be characterized with complete understanding of repeatability and variability.
6. There should be process ownership by everyone involved in the management process.
7. Management philosophy must be that no level of defect is acceptable.
8. Quality improvements from key process control actions must be documented and rewarded on a timely basis.
9. The management information system for process control must provide a real-time feedback mechanism to everyone involved in the quality improvement effort.
10. There should be provision for ongoing training in process control techniques.
11. The attainment of a good quality product should be through the prevention of defects, not through the detection of defects.
12. The removal of defects from a process should be done through root cause analysis and the implementation of proper corrective action. Costly inspections should be discouraged. The strategy should utilize sampling inspection when necessary and no inspection when not warranted.

5.4 BENEFITS OF THE PROCESS CONTROL MANAGEMENT SYSTEM

The competitive business environment demands a better quality product or service at the existing price or at a lower price. As a result, quality and productivity improvement through process control has emerged as a major new business strategy in many organizations. The strong emphasis on quality is driven by a number of reasons, including the following.

1. The acceptance of products produced or services rendered relies on conformance to requirements or specifications, which means all output produced must be defect free.
2. Quality improvement increases productivity.
3. The cost of production or services is increasing.
4. Product liability may lead to a loss in the market share.
5. Consumer education and awareness of quality are increasing.
6. There is intensive global competition in all industrial sectors.

In order for organizations to effectively offer good quality products and services that will compete in the world market, it is necessary for them to institute a statistical process control system. Such a system can have the following benefits:

1. Facilitates process capability improvement. It helps in determining the root cause of a problem and monitoring of corrective actions
2. Facilitates increases in productivity through control of rework, machine downtime, scrap, and work-in-process inventory
3. Provides a basis for both management and employee to understand and control assignable causes that affect a process performance
4. Facilitates the minimization of the "cost of quality," the expense of doing things right the first time
5. Creates a basis for effective control and understanding of the complex interaction among production and human variables, such as machines, fixtures, materials, and human work habits and efficiency
6. Facilitates improvements in quality yield and reduction in product cycle time
7. With tools, such as control charts, provides a common language for communications about the performance of a process

5.5 IMPLEMENTATION STEPS FOR STATISTICAL PROCESS CONTROL

The implementation of statistical process control concepts and techniques requires good control tools for performance and process adherence, clear understanding of supplier specifications, a timely reaction to problems within the process, a dedicated team effort to address process problems on a timely basis, and management commitment to quality and productivity improvement through process control. Each step for implementing statistical process control shown in Figure 5.2 will now be described.

5.5.1 Familiarization Sessions with Personnel, Processes, Technologies, Supplier Specifications, and Consumer Requirements

In this step, a formal study and understanding of the production or service processes, procedure, information system, data collection system, and product mix with associated technologies should be obtained. All input to the process and detailed understanding of supplier specifications should be obtained. A team of experts in statistical control concepts should review all the specifications and streamline them for the specified processes. A meeting should be held to introduce the purpose of the statistical process control in the work environment. It is usually helpful to demonstrate management commitment through participation in the kickoff meeting or through a memorandum of support. The key elements involved in this step are shown schematically in Figure 5.3.

5.5.2 Classification of Production or Service Process into Sector Cells

The production or service processes should be divided into sector cells. The cells should have the property of being able to process a family of products or services that share at least one common property, design or production or both. Other factors to consider are the work flow pattern, the source of delivery of raw materials, the ability to obtain a balanced process through a line balancing approach, and the interdependencies among variables and sector cells. The elements involved in this step are shown schematically in Figure 5.4.

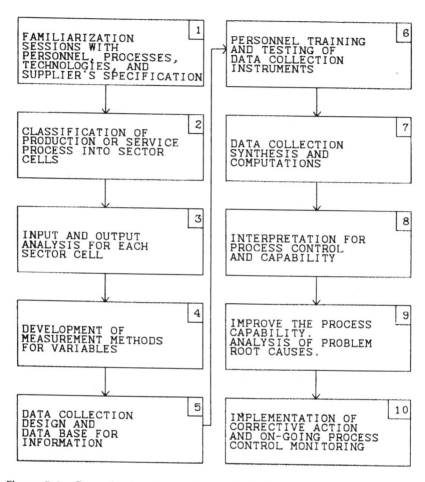

Figure 5.2   Steps for implementing statistical process control.

5.5.3   Input and Output Analysis for Each Sector Cell

In order to understand the characteristics of all production and service variables, input and output analysis of each cell is performed. The significance of each task performed and the characterization of both equipments and processes are obtained. For example, equipment

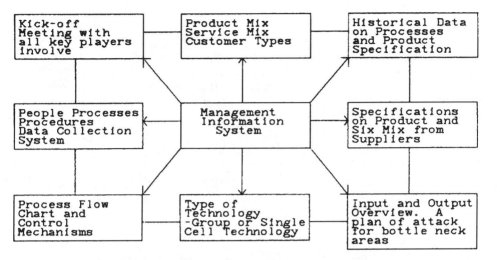

Figure 5.3   Elements involved in statistical process control familiarization sessions.

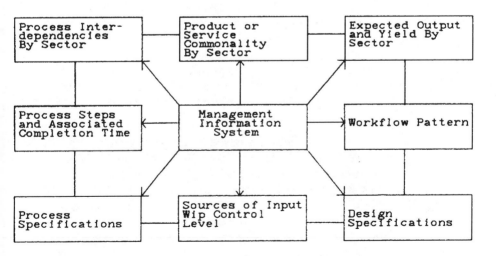

Figure 5.4   Elements involved in sector cell classification.

characterization should consider such items as variables that affect machine output, machine capability (uptime and downtime of machine), output per hour, and repeatability of machine problems. Process characterization should consider such items as the incoming quality of products, the impact of multiple products on one machine, process yields, and human/machine interface problems. The emphasis should be to determine the variables that affect process output and their effects on each sector cell and the total combination of cells. This will enable us to understand how each variable in each cell can be controlled. In situations in which the sector has a strong reliance on perishable items, the total time for each perishable good to become waste should be obtained. The elements involved in this step are shown schematically in Figure 5.5.

5.5.4 Development of a Measurement Method for Variables

The primary emphasis should be on the use of control charts in the understanding and reduction of fluctuations in a process until they are in a state of statistical control at the level desired.

Control Chart Definition

Control charts are a statistical device used for the analysis and control of repetitive processes. They are used to study variations in a process that are attributed to special causes. There are two types of control charts used to study variation in a process.

Attribute Control Charts. Control charts measure whether the product is defective based on a number of quality characteristics. The attribute control chart can be classified into different types; some examples are discussed below.

P chart (percentage defective) is based on attribute data (number of defective units of product). It is used to control the overall defective fraction of a process. The P chart provides an overall picture of quality and is often based on data from inspection records. However, it does not provide sufficient information for the control of individual characteristics. It does not provide information on the different degrees of defectiveness in the units of a product.

C chart (number of defects) is based on attribute data (number of

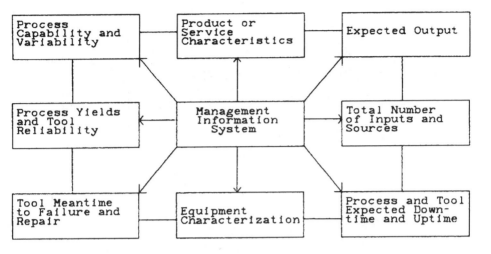

Figure 5.5   Elements involved in input-output analysis.

defects per unit of product). It is used to control the overall number of defects per unit. It provides a measure of defectiveness based on data from inspection records. However, it does not provide sufficient information for the control of individual characteristics.

The measurement methods for the various types of attribute control charts are presented in Table 5.1.

Variable Control Charts. These are control charts that measure the degree in variation of a single quality characteristic, such as tolerances or temperature. The X chart shows variations in the averages of samples. R charts show variations in the range of samples. Both the average (X) and range (R) charts are primarily used to control individual characteristics. These charts cannot be used with go or no go data. These charts are helpful when detailed information on the process average and variations are needed for the control of individual dimensions. Examples of the sample size usually used for each type of control chart are specified in Table 5.2. The work forms for both variable and attribute control charts are shown in Appendix B.

Table 5.1 Measurement Method for Attribute Control Charts

What is measured	Control chart	Sample size	What is to be controlled	Center line	Control limits	Comments
Number of defectives in sample	np chart	Constant	d: number of defectives in a constant sample size	$n\bar{p}$	$n\bar{p} \pm 3\sqrt{n\bar{P}(1-\bar{P})}$	n = sample size; $p = \dfrac{\text{number defectives}}{n}$; $\bar{p} = \dfrac{\text{total defectives}}{\text{total inspected}}$
Average number of defectives in sample	P chart	Varies	P: ratio of defectives to sample size	$\bar{p}$	$\bar{p} \pm \sqrt{\dfrac{\bar{P}(1-\bar{P})}{n}}$	
Number of defects in sample	C chart	Constant	c: number of defects in a constant sample size	$\bar{c}$	$\bar{c} \pm 3\sqrt{\bar{c}}$	$c = \dfrac{\text{total number defects}}{\text{number of samples}}$; More than one defect may be recorded on a piece in the sample
Average number of defects in sample	U chart	Varies	u: The ratio of nonconformities to sample size $u = \dfrac{c}{n}$	$\bar{u}$	$\bar{c} \pm 3\sqrt{\dfrac{\bar{u}}{n}}$	$c = \dfrac{\text{number of defects}}{\text{number of defects}}$; $u = \dfrac{\text{number of defects}}{\text{sample size}}$; $\bar{u} = \dfrac{\text{total number of defects}}{\text{total pieces inspected}}$

Table 5.2 Sample Size for Attribute and Variable Control Charts

	Average $\bar{X}$ and range R	% Defective (P)	Number of defects (C)
Sample size	Usually 4 or 5	Samples of 25, 50, or 100 from inspection results	Unit or product; example, one printed circuit board

5.5.5 Data Collection Design and Data Base for Statistical Process Control Information

In this step various forms and instruments are designed to capture the input and output components needed for statistical process control in each sector cell. In designing these forms, various factors, such as frequency of data collection, information required, computation steps, scale of measurement, and clarity, should be considered. An example of a measurement form for a variable control chart calculation is shown in Table 5.3. Blank work forms for data collection are shown in Appendix B. A data base should be set up for the storage and retrieval of information on process performance at each sector cell. Personal computers are useful systems for data base management.

5.5.6 Personnel Training and Testing of Data Collection Instruments

In this step, personnel associated with the implementation of statistical process control in each sector cell should be trained. An example is shown in Figure 5.6.

5.5.7 Data Collection, Synthesis, and Computations

In this step, the various forms designed in Section 5.5.5 are used to collect process yields, input and output components, and other measures for each sector cell. The data are collected periodically and synthesized and then used to compute the control limits and process averages for each sector cell, product type, and specific quality charac-

Table 5.3   Calculation Work Sheet for Variable Control Chart[a]

Calculate average range ($\sum$ = sum of, and k = number of subgroups)

$$\bar{R} = \frac{\sum R}{k} = \underline{\hspace{1cm}} = \underline{\hspace{1cm}}$$

Calculate control limits for ranges:

$$UCL_R = D_4\bar{R} = \underline{\hspace{2cm}} \times \underline{\hspace{2cm}} = \underline{\hspace{2cm}}$$

$$LCL_R = D_3\bar{R} = \underline{\hspace{2cm}} \times \underline{\hspace{2cm}} = \underline{\hspace{2cm}}$$

Calculate Grand Average:

$$\bar{\bar{X}} = \frac{\sum \bar{X}}{k} = \underline{\hspace{1cm}} = \underline{\hspace{1cm}}$$

Calculate control limits for ranges:

$$A_2\bar{R} = \underline{\hspace{2cm}} \times \underline{\hspace{2cm}} = \underline{\hspace{2cm}}$$

$$UCL_{\bar{X}} = \bar{\bar{X}} + A_2\bar{R} = \underline{\hspace{2cm}} + \underline{\hspace{2cm}} = \underline{\hspace{2cm}}$$

$$LCL_{\bar{X}} = \bar{\bar{X}} - A_2\bar{R} = \underline{\hspace{2cm}} - \underline{\hspace{2cm}} = \underline{\hspace{2cm}}$$

Estimate of standard deviation (if the process is in statistical control):

$$\sigma = \frac{\bar{R}}{d_2} = \underline{\hspace{2cm}}$$

[a]Control limits based on subgroups.

teristic of interest. The elements involved in this step are shown schematically in Figure 5.7.

5.5.8   Interpretation for Process Control and Capability

The process yields (defects, process averages, upper control limit, and lower control limit) of each sector cell are analyzed periodically to determine why they increased or decreased. Emphasis should be placed on understanding the pattern of variation and pinpointing assignable causes for the variation within the process.

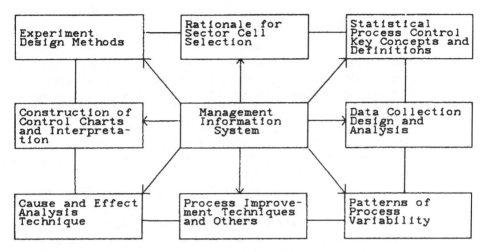

Figure 5.6   Elements to be considered in statistical process control training.

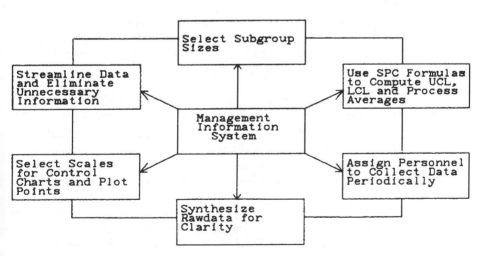

Figure 5.7   Elements involved in data collection, synthesis, and computations.

Understanding Process Patterns of Variation

As a process of any system operates over a period of time, typical patterns develop. It is important for the process control analyst to recognize these patterns for better interpretation and analysis of process problem root causes. The fourteen most common types of patterns are now presented.

Trend. A cumulative trend usually consisting of seven or more consecutive points indicates that the process is drifting. The pattern may be a gradual rise or fall to either portion of the control chart. The possible root cause of the trend may be tool wear, depletion of reagent, or a decline in human efficiency owing to fatigue. This pattern is shown schematically in Figure 5.8.

Lack of Stability. This pattern reflects a high level of large fluctuations within the process. The possible root causes of the fluctuations may be overadjustment of process parameters, such as material, tools, and equipment, and overcontrol of the process. This pattern is shown schematically in Figure 5.9.

Sudden Shift in Level. A qualified process will experience a sudden shift in level if new equipment, new operators, or new materials are introduced to the process. A difference in product mix and configuration may also shift the process level. This pattern is shown schematically in Figure 5.10.

Points Warning. When points are outside the upper or lower control limit, a special assignable cause is often responsible, for example, a high level of machine misinsertion within a given time period or a lack of operator training for a given task. This pattern is shown schematically in Figure 5.11.

Cycles. Differences among work shifts, operators, temperature, and other factors may result in a consistent pattern of repeated low and high points that occurs periodically almost within the same time intervals. This pattern is shown schematically in Figure 5.12.

Normal. A normal pattern occurs within the control limits and is usually based on random variation. This pattern is shown schematically in Figure 5.13.

Occasional Freaks. This are points that occur outside the control limits with no consistent pattern. Freaks are usually caused by the interaction of multiple factors, which is difficult to analyze. The best source for pinpointing the root causes for freaks is in the design of the experiment. This pattern is shown schematically in Figure 5.14.

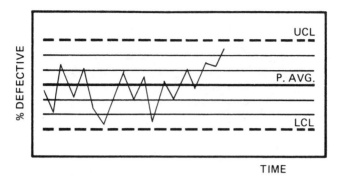

Figure 5.8   A typical trend pattern in statistical process control.

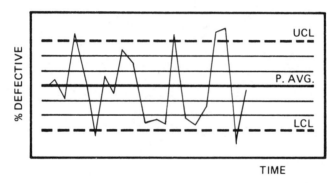

Figure 5.9   Lack of stability pattern in statistical process control.

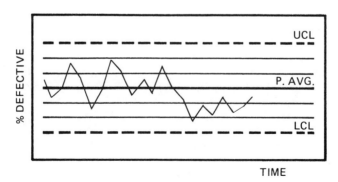

Figure 5.10   Sudden shift in level pattern in statistical process control.

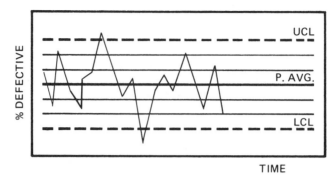

Figure 5.11   Points warning pattern in statistical process control.

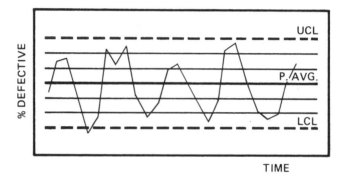

Figure 5.12   Cyclic pattern in statistical process control.

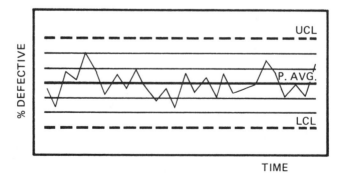

Figure 5.13   Normal pattern in statistical process control.

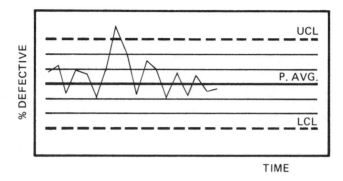

TIME

Figure 5.14   Occasional freaks pattern in statistical process control.

Natural Pattern. The process is fairly stable with no significant variation. The state of control is maintained consistently over time. This pattern is shown schematically in Figure 5.15.

Runs and No Runs. This pattern indicates too many points or too few points on one side of a control chart centerline. This pattern could be caused by using the wrong specification for production, such as wrong equipment parts. An example of this pattern is shown Figure 5.16.

Predictable Change. This pattern shows a point-to-point variation that is easily identified. The variation is usually caused by a similar or the same cause, such as the level of tool wear. An example of this pattern is shown in Figure 5.17.

Stratification. These are unnaturally small fluctuations that often appear around the centerline of control charts. Figure 5.18 shows this type of pattern.

Mixtures. A few points near the centerline of the control chart show a seesaw effect. This pattern is usually caused by differences among sources of material and types, different training levels between operators, multiple machines, multiple tools, and two different temperatures. This pattern is shown schematically in Figure 5.19.

Cyclic Shift in Level. This pattern occurs when items, such as new material, are introduced to the process at the same time internally or seasonally. A seasonal variation in humidity and temperature levels as it affects tools and equipment within the process can also cause this pattern, shown in Figure 5.20.

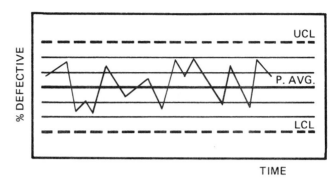

Figure 5.15   Natural pattern in statistical process control.

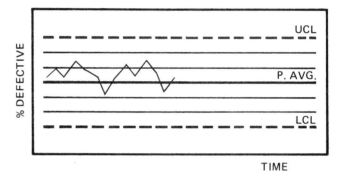

Figure 5.16   Runs and no runs pattern in statistical process control.

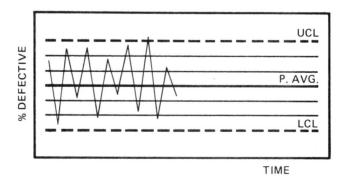

Figure 5.17   Predictable change pattern in statistical process control.

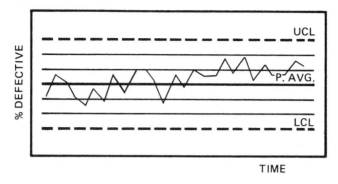

Figure 5.18 Stratification pattern in statistical process control.

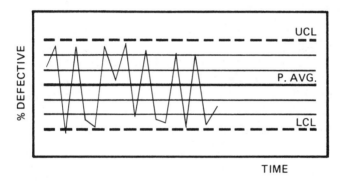

Figure 5.19 Mixed pattern in statistical process control.

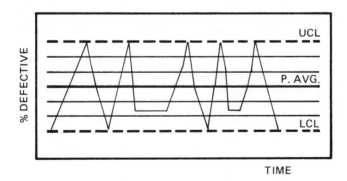

Figure 5.20 Cyclic shift in level pattern in statistical process control.

Systematic Variables Skip Repeat. This pattern is characterized by a predictable point-to-point variation that occurs within a given time internal. The predictable point-to-point variation often follows a different steady-state level in the subsequent time periods. This pattern is shown schematically in Figure 5.21.

Evaluating Process Runs

In order to properly evaluate runs, the control chart is divided into zones. Each zone is 1 standard deviation wide. Between the upper and lower control limit are a total of 6 standard deviations. The centerline shown in Figure 5.22 divides the control chart into two equal zones of 3 standard deviations each. As stated by Juran and Gryna (1980), process instability is determined by the following:

1. A single point falls out 3 standard deviation limits (beyond zone A).
2. Two of three successive points fall in zone A or beyond (the odd point may be anywhere).
3. Four of five successive points fall in zone B or beyond (the odd point may be anywhere).
4. Eight successive points fall in zone C or beyond.

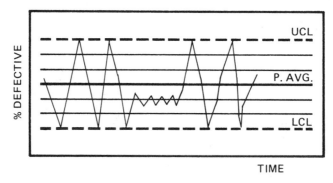

Figure 5.21   Systematic variable skip repeat pattern in statistical process control.

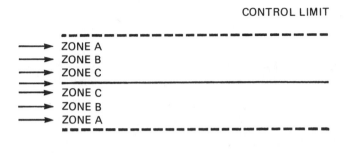

Figure 5.22   Zones for applying tests for instability. (Adapted from Juran, J. M., and F. M. Gryna, 1980. *Quality Planning and Analysis*, McGraw-Hill, New York, p. 340. Reprinted with permission.)

Evaluating the Process Capability

The most generally used approach for evaluating the capability of a process to satisfy customer requirements in the C ± 3σ capability, where C is the process average (centerline) of the control chart. C ± 4σ is often used for new processes. These requirements are often based on a minimum performance level that has relative consistency within a product mix and other quality characteristics. There are situations in which the producer may wish to tighten the process capability requirements in order to satisfy a particular customer's demand. However, it is recommended that a trade-off analysis be conducted before changing the capability limits for an entire production output.

5.5.9   Improve Process Capability Through Analysis of Problem Root Causes

In this step the sector cell team members meet regularly to analyze the patterns of the process based on the data collected. The team focuses on assignable causes that are responsible for process performance. The root causes of problems are sought and improvement actions implemented to correct problems. It is recommended that when new improvement actions are implemented, they should be tracked carefully to monitor the impact on process performance. The elements involved in this step are shown schematically in Figure 5.23.

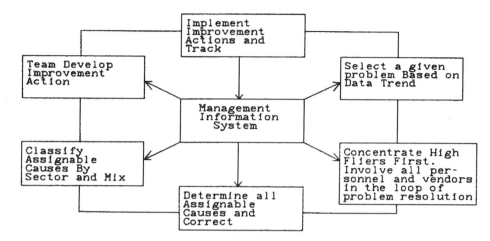

Figure 5.23   Elements involved in process analysis for problem root causes.

5.5.10   Ongoing Process Control Monitoring

Statistical process control requires an ongoing process that emphasizes quality at the source of production. The prevention system for defects is necessary. All key actions implemented within the process must follow a means of de-emphasizing the detection and reaction system. Corrective action strategy requires total team involvement and commitment. There should be assigned tasks with responsibility and accountability. The key elements involved in this step are presented in Figure 5.24.

5.6   BASIC CONCEPTS FOR DESIGNING EXPERIMENTS IN THE BUSINESS PROCESS ENVIRONMENT

It is often difficult in certain situations to be able to pinpoint the causes of variation within a given process because of abnormal conditions that may be the result of a number of different factors. Experiments are carried out to achieve the following:

1. Provide information and facts needed for the prediction of a process behavior and performance

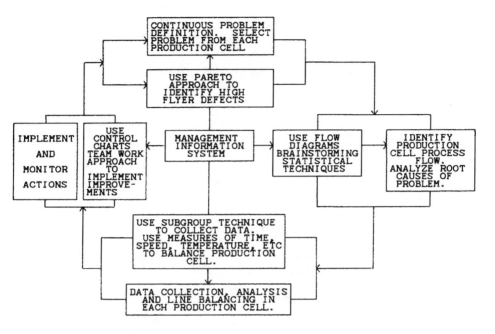

Figure 5.24 The key elements in ongoing process control management.

2. Discover something about controllable factors and uncontrollable variables in a particular process
3. Compare the effect of several factors on some phenomena
4. Identify root causes of assignable causes of given problem within a process
5. Provide information needed for the prediction of an improvement action or process change within a given process

5.6.1 Three Basic Principles of Experimental Design

The three basic principles of experimental design are as follows.

Replication involves repetition of a basic experiment. The experimental error can be estimated, and a sample mean can be a useful tool to obtain a precise estimate of parameters.

Randomization involves the allocation of the experimental material and the order in which an individual runs the experiment. These are randomly determined.

Blocking involves making comparisons among the conditions of interest in the experiment within each block. It is used to increase the precision of an experiment.

The key principles and terminology associated with experimental design are shown schematically in Figure 5.25.

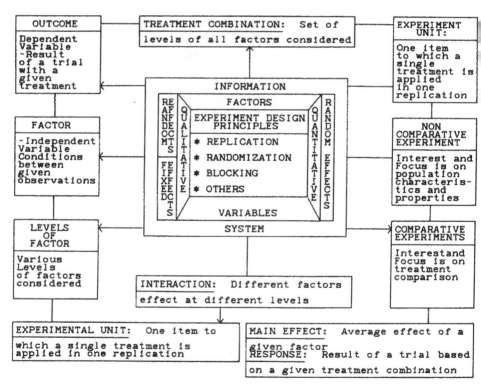

Figure 5.25   Principles and terminology in experimental design.

5.6.2 Requirements for a Successful Experiment

As discussed in Section 5.5, understanding the assignable causes for a process pattern is not any easy task. Experiments that aid in the interpretation and understanding of the root cause problems of a process must be based on the following. .

1. The experiment must be based on an efficient design that recognizes all process variables and parameters.
2. The experiment must process no systematic error in the execution, design, and implementation phases.
3. The experiment must use homogeneous materials and have control mechanisms in place for extraneous factors.
4. The experiment must have an acceptable experimental error, and the results obtained must be statistically significant.
5. The conclusions drawn from the results of the experiment must have a wide range of validity and application.
6. The data obtained from the results of the experiment must lend themselves to statistical analysis.
7. The controlled environment of the experiment must be clearly specified and all inferences drawn must be limited to this or a similar environment only.

5.6.3 Steps in Designing a Process Experiment

A ten-step approach for designing experiments is specified in Figure 5.26. Each step will now be described.

Step 1. The process problem and area of impact are identified. Historical evidence of the existence of a problem or estimate of trends to support the definition of a problem must be provided. Other interrelated variables that need further clarification must also be specified.

Step 2. The specific problem statement is translated to a null and alternative hypotheses to enable the data obtained to be analyzed statistically.

Step 3. The planning and implementation of the experiment will be successful if all personnel involved in the problem area of impact have involvement in specifying variables, factors, and parameters. This provides a clear idea of what is studied, how all relevant information is obtained, and the usefulness of the result of the experiment. The resources

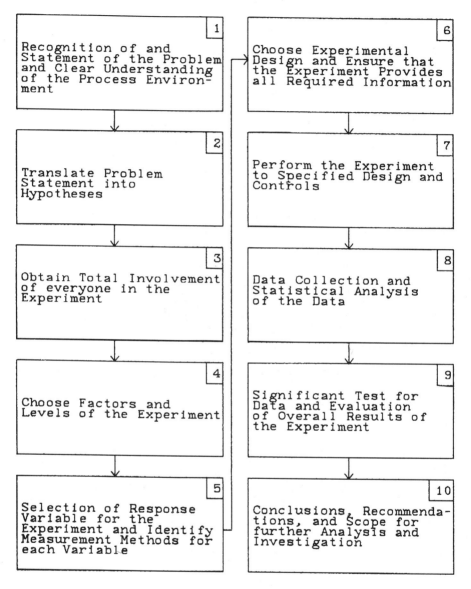

Figure 5.26   Steps in a designing process experiment.

needed to perform and implement the results of the experiment are also easily obtained through cooperation from everyone involved.

Step 4. In this step a distinction is made between quantitative and qualitative factors, dependent and independent variables, and the range over which all factors and variables are selected. Other parameters, such as the number of levels of runs and factor interaction, are selected.

Step 5. All dependent variables are clearly specified. Dependent variables and other variables must be selected based on facts, not opinions. The measurement methods for each variable should be specified.

Step 6. It is important to note that complicated experiments are expensive in terms of both increased possibility of error and increased cost. It is recommended that experimental design be keep as simple as possible. The choice of experimental design must be based on the following factors: cost, time, risk, accuracy level, and the true response range of all resources needed for the detection of causes assignable to the problem being investigated.

Step 7. In this step, particular attention must be paid to randomization, measurement accuracy, and maintaining as uniform an experimental environment as possible. Appropriate measurement instruments are designed for data collection with a random check on the validity of data source and frequency.

Step 8. Data should be collected periodically based on the specified time table for the experiment. The data should be synthesized and all unnecessary information weeded out.

Step 9. In this step, statistical techniques, such as the analysis of variance (ANOVA), and other simple tools, such as mean and variances, are used to analyzed the data collected from the experiment. The techniques must be able to specify the range of validity or significance of the results obtained from the experiment. The statistical technique used must be able to show a confidence interval for results. Computer programs and graphic analysis techniques should be employed when applicable to eliminate tedious manual computations.

Step 10. The conclusions and recommendations from the experiment must be drawn from the results obtained. The practical significance and implication of the results and findings must be clearly stated. Recommended actions to resolve the problem investigated must be specified. The exposures and limitations of findings and recommendations must be pointed out to avoid further problems when solutions are implemented. The scope for further investigation should be identified to enable continued further analysis of improvement in the process. Managerial judgment is necessary to make trade-offs in situations that involve risk, accuracy level, cost factors, and other variables.

5.7 CASE STUDY: STATISTICAL PROCESS CONTROL IN A GROUP TECHNOLOGY ENVIRONMENT*

The statistical process control implementation steps presented in Section 5.5 were applied in a manufacturing plant that produces printed circuit boards to customer order. The entire production process from raw materials to finished goods was divided into four group technology cells. The manufacturing process methods were revised to enable complete cell balance based on production parameters.

For each cell, control charts and measures were developed to monitor the process performance and yield. A computer data base was developed for each cell. Within each cell both process and system enhancement was made. The results obtained after the implementation of statistical process control in each cell is shown in Figures 5.27 through 5.30. In addition to the significant improvement in quality in each group technology cell, it was observed that partial productivity (labor productivity) in each cell increased significantly after the implementation of statistical process control.

Other benefits obtained were a reduction in cycle time and inventory, improvement in the percentage of goods conforming printed circuit boards, less rework, and a reduction in setup time. The production planning process for the entire group technology environment also became easier after the implementation of statistical process control. Statistical process control made possible a clearer understanding and prediction of the parameters and variables in each production cell.

5.7.1 Problems Encountered During Implementation

The problems presented here are mainly intended to form a guide for companies implementing statistical process control in a group technol-

---

*Adapted from Edosomwan, 1986.

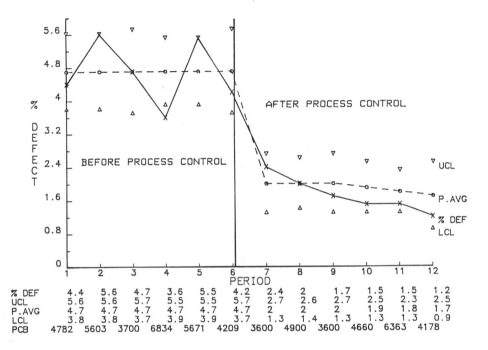

Figure 5.27   Statistical process control result from group technology cell 1.

ogy environment. It must be pointed out that problems associated with statistical process control in a GT environment could vary depending on the organizational setting and production environment. Based on the case study presented, we found that incoming product quality posed a significant problem at the machine level. This problem was resolved by characterization of products by vendors and front-end screening before these products are allowed in each GT cell.

The vendors were made to take responsibility for the correction of incoming defects. Grouping the processes, machines, products, and other factors into group technology (GT) cells would be a tremendous task. To overcome this problem, a teamwork approach was used in the cell balancing requirements. This also enabled any resistance to the study to fade away. Data collection, compilation, and computation were also found to be cumbersome. A computer program was developed for this purpose that eased the data management burden. The characterization

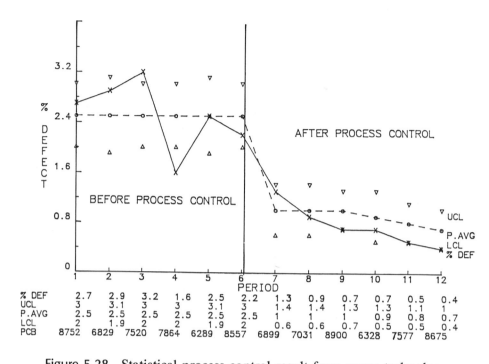

Figure 5.28   Statistical process control result from group technology cell 2.

of each GT cell required a flexible work schedule driven by high customer demands for existing products. This problem was overcome through a rotational work schedule among people and departments. Initially, a crisis developed when a problem in a particular GT developed without a clear understanding of the cause or of who and what was responsible. To overcome this problem, specific tasks were assigned to specific people and meetings were held regularly to discuss and resolve problems in real time. Training was provided for both management and employees to facilitate good root cause analysis for problems and implementation of corrective actions.

The group technology approach to production management is a technique that will become increasingly important in the era of

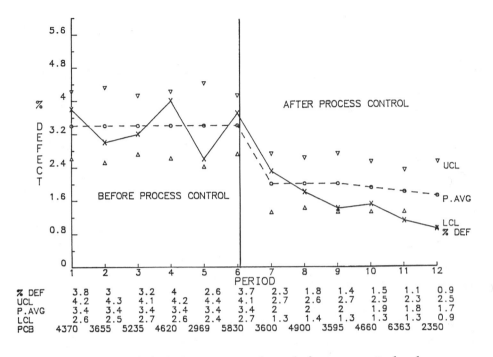

Figure 5.29   Statistical process control result from group technology cell 3.

technological explosion. GT by itself does not offer a greater advantage than other traditional approaches. However, with the implementation of statistical process control in the GT environment, tremendous benefits and potentials exist in productivity and quality improvement. The concept of statistical process control in a GT environment is also still in its early stages of testing and implementation. More work is still needed to provide a systems-based approach for controlling and managing the parameters within each GT cell. The case study followed an approach that has a wide range of applications in various GT groups, including CAD/CAM robotics. Success in the use of the approach suggested depends in part on management and employee commitment to excellence through quality and productivity improvement at the source of production.

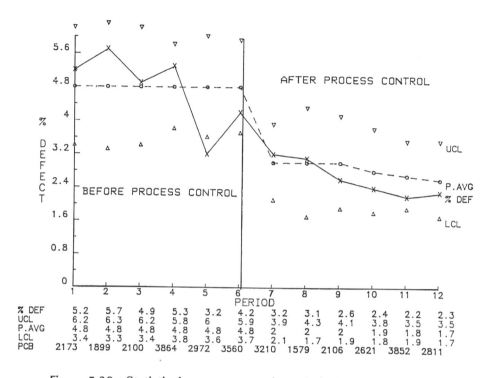

% DEF	5.2	5.7	4.9	5.3	3.2	4.2	3.2	3.1	2.6	2.4	2.2	2.3
UCL	6.2	6.3	6.2	5.8	6	5.9	3.9	4.3	4.1	3.8	3.5	3.5
P.AVG	4.8	4.8	4.8	4.8	4.8	4.8	2	2	2	1.9	1.8	1.7
LCL	3.4	3.3	3.4	3.8	3.6	3.7	2.1	1.7	1.9	1.8	1.9	1.7
PCB	2173	1899	2100	3864	2972	3560	3210	1579	2106	2621	3852	2811

Figure 5.30   Statistical process control result from group technology cell 4.

5.8 BASIC STATISTICAL TOOLS FOR DATA ANALYSIS

Statistics provides useful scientific procedures for collecting, describing, organizing, summarizing, and analyzing quantitative information on any type of problem that may be encountered within any economic unit. Additional benefits that may be obtained from statistical analysis are shown schematically in Figure 5.31. Following this, some of the most commonly used statistical tools are presented.

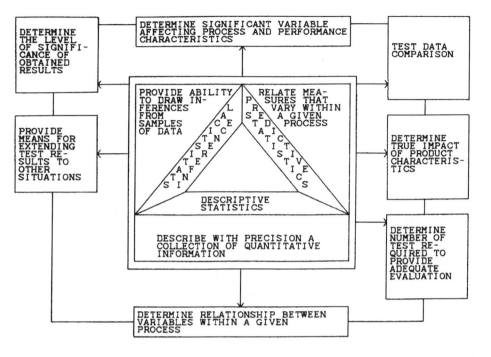

Figure 5.31 Benefits of statistical analysis.

5.8.1 Frequency Distributions

A frequency distribution is a technique that enables the adequate description of data based on the frequency of occurrence of the different values of the variables. There are two types of frequency distribution: (1) a frequency distribution based on ungrouped data, and (2) a frequency distribution based on grouped data. Both distributions have three major characteristics, as shown in Figure 5.32: central value characteristics, shape, and spread.

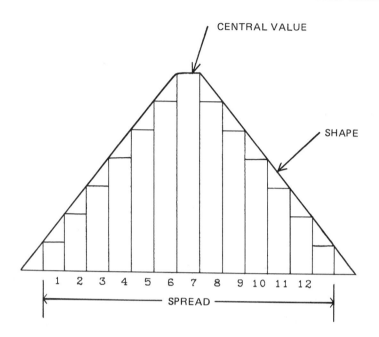

Figure 5.32   Characteristics of a frequency distribution.

Frequency Notations and Formulas

Let

X = all possible data values within a class interval (grouped data)
or within an absolute number (ungrouped data)

M = mean of data in a frequency distribution

f = frequency of data occurrence within the specified range

R_f = relative frequencies based on the total number of values in
the collection

N = total number of values in the collection

P_r = percentile rank of any value in a frequency distribution with
unit interval

C_f = cumulative frequency of a given interval (frequency of the interval plus the total of the frequencies of all intervals below the
given interval)

R_{cf} = relative cumulative frequency

$$R_f = \frac{f}{N} \tag{5.1}$$

$$R_{cf} = \frac{C_f}{N} \tag{5.2}$$

$$P_r = \frac{100}{N} \left(C_f - \frac{f}{2} \right) \tag{5.3}$$

$$M = \frac{E_{fi}X_i}{N} \tag{5.4}$$

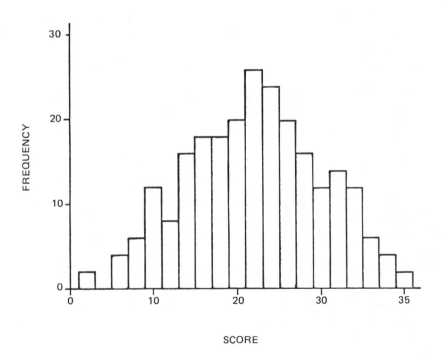

Figure 5.33   Histogram (frequencies represented in the form of vertical bars).

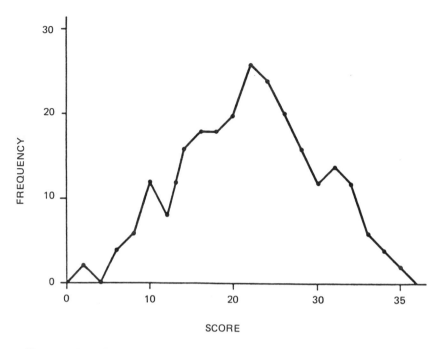

Figure 5.34   Frequency polygon (frequencies represented in straight lines connecting points).

Graphs Representing Frequency Distributions

A graphic technique is useful in showing the spread of frequency distributions. The technique enables a pictorial look that aids in the analysis of data collected. The various types of graphic techniques are shown schematically in Figures 5.33 through 5.38.

5.8.2   Arithmetic Mean

The mean, often referred to as the average, is the most commonly used measure of central tendency. Let

X_i = value of any interval (i = 1, 2, 3, . . . , N)

N = total number of values in the collection

$\bar{X}$ = mean or average of the population

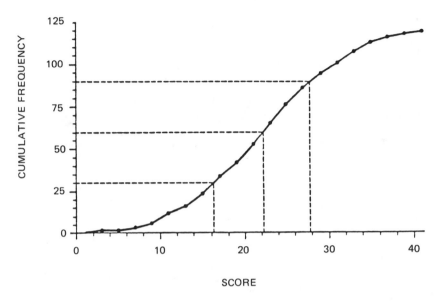

Figure 5.35 OGIVE (vertical scale represented in either relative frequencies, percentile ranks, or cumulative frequency).

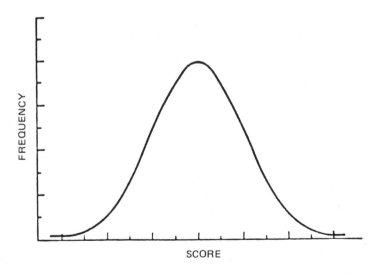

Figure 5.36 Normal curve (frequencies represented as continuous, symmetrical, bell-shaped curve).

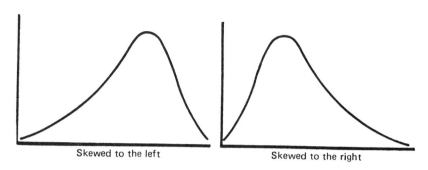

Figure 5.37   Skewness (symmetry or a symmetry of a curve).

$$\bar{X} = \frac{X_1 + X_2 + X_3 + X_n}{N} = \frac{\sum\limits_{i=1}^{N} X_i}{N} \qquad (5.5)$$

5.8.3   Median

Once data are ranked, the median is the middle observation. It is most commonly used for ordinal measures. It lies at the

$$\frac{(N + 1)\text{th}}{2} \qquad \text{observation} \qquad (5.6)$$

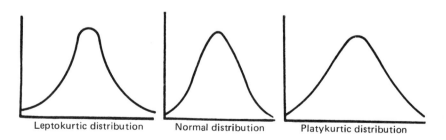

Figure 5.38   Kurtosis (relative peakedness or flatness of a curve).

5.8.4 Midrange

The midrange M_r of data is based on the maximum and minimum value within the data set.

$$M_r = \frac{X_{i\,max} + X_{i\,min}}{2} \tag{5.7}$$

5.8.5 Mode

The mode M_d measures the values that occur most often in a given data set.

$$M_d = \text{most common value } X_i - \text{most frequent occurrence } X_i \tag{5.8}$$

5.8.6 Range

The range R is the difference between the largest and smallest values in the collection (data set).

$$R = X_{i\,max} - X_{i\,min} \tag{5.9}$$

5.8.7 Proportion

The proportion P is often expressed as a percentage. It represents the total percentage of values with the same characteristics, such as defect levels within a given process.

$$P = \frac{X}{N} \tag{5.10}$$

where X = number of values having the same designated property.

5.8.8 Variance and Standard Deviation

The variance is the mean of the squared deviations from the mean and is usually abbreviated as mean square. The sample variance is used to estimate the population variance of a collection (data set). The standard deviation is simply the square root of the variance. Let S = standard deviation.

$$S = \sqrt{\frac{\sum\limits_{i=1}^{N} (X_i - \bar{X})^2}{n-1}} \qquad \sqrt{\frac{n \sum X^2 - \left(\sum X\right)^2}{n(n-1)}} \tag{5.11}$$

Variance $= S^2 \tag{5.12a}$

The coefficient of variation (standard deviation as a percentage of the mean V) is expressed as

$$V = \frac{S}{\bar{X}} 100 \tag{5.12b}$$

5.9 BASIC REVIEW OF PROBABILITY

The probability measure is the quantification of the likelihood that a particular event will occur. The probability rules are derived from set theory.* Figure 5.39 shows an example of the universal set containing all objects within which subobjects exist.

5.9.1 The Addition Rule

Let

   $P(A)$ = probability that event A will occur

   $P(B)$ = probability that event B will occur

   $P(C)$ = probability that event C will occur

   $P(1)$ = 1 (certainty)

$$P(A \text{ or } B) = P(A) + P(B) - P(AB) \tag{5.13}$$

$$P(P \text{ u } B) = P(A) + P(B) - P(AB) \tag{5.14}$$

$$P(A \text{ or } B \text{ or } C) = P(A) + P(B) + P(C) - P(AB) - P(BC)$$
$$- P(AC) + P(ABC) \tag{5.15}$$

5.9.2 Conditional Probability

If A and B are independent,

$$P\left(\frac{A}{B}\right) = \frac{P(A \cap B)}{P(B)} \tag{5.16}$$

where

$$P(A \cap B) = P(AB) \tag{5.17}$$

---

*Detailed theoretical discussion can be found in Duncan, 1974. *Quality Control and Industrial Statistics*, 4th ed. Richard D. Irwin, Homewood, Illinois.

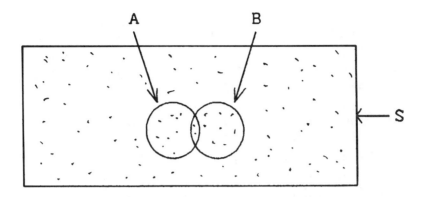

Figure 5.39   Universal set containing two objects.

$$P(ABC) = P(A) \, P(B/A) \, P(C/AB) \tag{5.18}$$

Suppose $A_1, A_2, \ldots, A_y$ are mutually executive of the one event that must occur; therefore,

$$P\left(\frac{A_i}{B}\right) = \frac{P(A_i) \, P(B/A_i)}{\sum P(A_i) \, P(B/A_i)} \tag{5.19}$$

5.9.3   Count Tools

Permutation of Objects

An ordered arrangement of r distinct objects is called permutation. The number of ways of ordering N distinct objects taken r at a time is designated by

$$P_r^N = \frac{N!}{(N - r)!} \tag{5.20}$$

Expression (5.20) is referred to as the Bays theorem.

Combination of Objects

The number of combinations of N objects taken r at a time is the number of subsets, each of size r, that can be formed from N objects and is designated

$$C_r^N = \frac{N!}{(r!)(N - r)!} \qquad (5.21)$$

5.10 CONFIDENCE INTERVAL, LIMITS LEVEL, AND POINT ESTIMATE

The confidence interval is a range of values that include the true value of a collection (data set of a population).

Confidence limits are the upper and lower boundaries of the confidence interval.

The confidence level is the value of the preassigned probability used to determine the confidence interval. Confidence levels of 90, 95, and 99 are most commonly used.

A point estimate is a single-valued statistic used for the determination of an unbiased estimate of a given parameter.

A confidence limits computation is primarily based on the population mean, preassigned probability, sample size, degrees of freedom, and a distribution. The formulas for computing confidence limits are specified in Table 5.4.

Degrees of freedom (DF) is considered when a sample standard deviation is used to estimate the true standard deviation of a universe. DF can be expressed as follows:

DF = number of measurements in the sample − number of
     constraints estimated from the data in order to compute
     standard deviation   (5.22)

The sampling distribution of the means shown in Figure 5.40 provides a basis for visualizing the total area under the curve at different standard deviations. The more commonly used intervals are 90, 95, and 99 percent.

5.11 HYPOTHESIS TESTING

A test of a hypothesis is done by analyzing a sample of data to validate a certain assertion. Two types of sampling errors can occur in evaluating a hypothesis.

Table 5.4 Confidence Limits Formulas

Parameters	Formulas
Mean of a normal population (standard deviation known)	$\bar{X} \pm Z_{\alpha/2} \dfrac{\sigma}{\sqrt{n}}$

where $\bar{X}$ = sample average
Z = normal distribution coefficient
σ = standard deviation of population
n = sample size

Mean of a normal population (standard deviation unknown)	$\bar{X} \pm t_{\alpha/2} \dfrac{s}{\sqrt{n}}$

where t = distribution coefficient (with $n - 1$ degrees of freedom)
s = estimated σ

Standard deviation of a normal population	Upper confidence limit = $s \sqrt{\dfrac{n-1}{\chi^2_{\alpha/2}}}$
	Lower confidence limit = $s \sqrt{\dfrac{n-1}{\chi^2_{1-\alpha/2}}}$

where χ^2 = chi-square distribution coefficient with $n - 1$ degrees of freedom
$1 - \alpha$ = confidence level

Difference between the means of two normal populations (standard deviations σ_1 and σ_2 known)	$(\bar{X}_1 - \bar{X}_2) \pm Z_{\alpha/2} \sqrt{\dfrac{\sigma_1{}^2}{n_1} + \dfrac{\sigma_2{}^2}{n_2}}$
Difference between the means of two normal populations ($\sigma_1 = \sigma_2$ but unknown)	$(\bar{X}_1 - \bar{X}_2) \pm t_{\alpha/2} \sqrt{\dfrac{1}{n_1} + \dfrac{1}{n_2}}$ $\times \sqrt{\dfrac{\sum (X - X_1)^2 + \sum (X - X_2)^2}{n_1 + n_2 - 2}}$
Mean time between failures based on an exponential population of time between failures	Upper confidence limit = $\dfrac{2rm}{\chi^2{}_{\alpha/2}}$

(continued)

Table 5.4   (*Continued*)

Parameters	Formulas

$$\text{Lower confidence limit} = \frac{2rm}{\chi^2_{1-\alpha/2}}$$

where r = number of occurrences in
the sample (i.e., number of
failures)
m = sample mean time
between failures
DF = 2r

1. Type I error. This sampling error occurs when a true hypothesis is rejected. The probability of a type I error is usually denoted by α.
2. Type II error. This sampling error occurs when a false hypothesis is accepted. The probability of a type II error is usually denoted by β.
3. Null and alternative hypotheses. The null hypothesis assumes that there is no difference between specified population param-

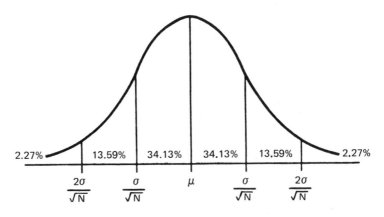

Figure 5.40   Sampling distribution of the means.

eters; the alternative hypothesis assumes that there is a difference between specified population parameters.

4. Possible results in hypothesis testing. The results in hypothesis testing can vary depending on the decision after analysis. The various possible outcomes are shown schematically in Figure 5.41.

5. Procedure for testing hypothesis. The procedure specified in Figure 5.42 is applicable to the testing of most statistical hypotheses.

6. Hypothesis testing formulas. The most commonly used statistical formulas for hypothesis testing are specified in Table 5.5

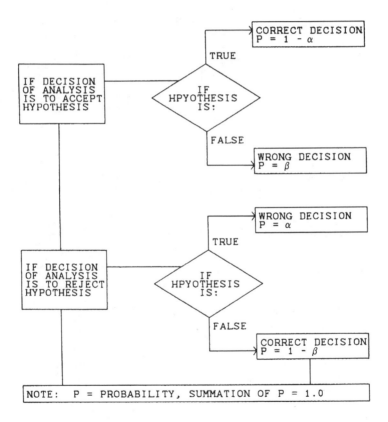

Figure 5.41   Possible results in hypothesis testing.

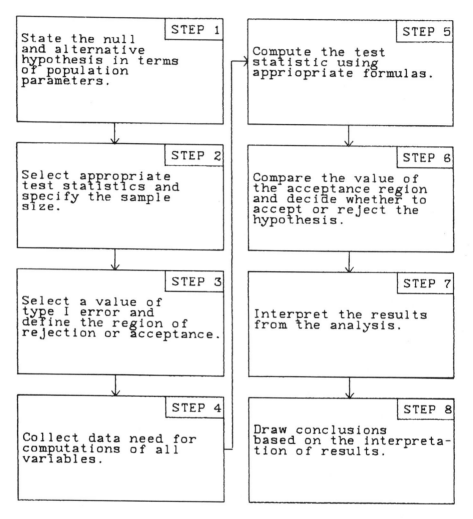

Figure 5.42   Steps in hypothesis testing.

Table 5.5　Formulas for Hypothesis Testing

Hypothesis	Test statistic and distribution
.H:$\mu = \mu_0$ (the mean of a normal population is equal to a specified value μ_0; σ is known)	$$Z = \frac{\bar{X} - \mu_0}{\sigma/\sqrt{n}}$$ Normal distribution
.H:$\mu = \mu_0$ (the mean of a normal population is equal to a specified value μ_0; σ is estimated by s)	$$t = \frac{\bar{X} - \mu_0}{s/\sqrt{n}}$$ t distribution with $n - 1$ degrees of freedom (DF)
.H:$\mu_1 = \mu_2$ (the mean of population 1 is equal to the mean of population 2; assume that $\sigma_1 = \sigma_2$ and that both populations are normal)	$$t = \frac{\bar{X}_1 - \bar{X}_2}{\sqrt{1/n_1 + 1/n_2}\sqrt{[(n_1 - 1)s_1^2}}$$ $$\overline{+ (n_2 - 1)s_2^2]/(n_1 + n_2 - 2)}$$ t distribution with DF $= n_1 + n_2 - 2$
.H:$\sigma = \sigma_0$ (the standard deviation of a normal population is equal to a specified value σ_0)	$$\chi^2 = \frac{(n - 1)s^2}{\sigma_0^2}$$ Chi-square distribution with DF $= n - 1$
.H:$\sigma_1 = \sigma_2$ (the standard deviation of population 1 is equal to the standard deviation of population 2; assume that both populations are normal)	$$F = \frac{s_1^2}{s_2^2}$$ F distribution with $DF_1 = n_1 - 1$ and $DF_2 = n_2 - 1$
.H:$p = p_0$ (the fraction defective in a population is equal to a specified value p_0; assume that $np_0 \geqslant 5$)	$$Z = \frac{p - p_0}{\sqrt{p_0(1 - p_0)/n}}$$ Normal distribution
.H:$p_1 = p_2$ (the fraction defective in population 1 is equal to the fraction defective in population 2; assume that n_1p_1 and n_2p_2 are each $\geqslant 5$)	$$Z = \frac{X_1/n_1 - X_2/n_2}{\sqrt{\hat{p}(1 - \hat{p})(1/n_1 + 1/n_2)}}$$ $$\hat{p} = \frac{X_1 + X_2}{n_1 + n_2}$$ Normal distribution

5.12 BASIC ANALYSIS OF VARIANCE TECHNIQUE

The basic analysis of variance technique (ANOVA) can be used to test the hypothesis that two or more independent samples drawn from the same population have the same mean. The procedure for performing the basic ANOVA is specified in Figure 5.43. Table 5.6 also shows a typical ANOVA matrix.

5.12.1 Formulas and Computational Procedure for Basic ANOVA

Symbols and Notations

Let

i = subscript for different level of factor A

j = subscript for different level of factor B

k = subscript for different cell values

a = number of levels for factor A

b = number of levels for factor B

n = number of observations per cell

X_{ijk} = cell values for each combination

T = correction term

SS_a = sum of squares for factor A

SS_b = sum of squares for factor B

SS_{ab} = sum of squares for interaction between A and B

SS_{tr} = sum of squares for between treatments

SS_e = sum of squares for error

MS_a = mean squares for factor A

MS_b = mean squares for factor B

MS_{ab} = mean squares for the interaction

MS_e = mean squares for error

HO_1 = first null hypothesis

HO_2 = second null hypothesis

HO_3 = third null hypothesis

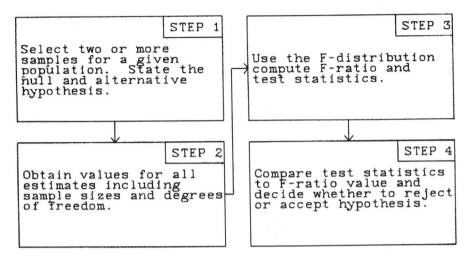

Figure 5.43   Steps in performing ANOVA.

$$T = \sum_i \sum_j \sum_k X_{ijk} \qquad (5.23)$$

$$SS_t = \sum_i \sum_j \sum_k X_{ijk}^2 - \frac{T^2}{nab} \qquad (5.24)$$

$$SS_a = \frac{\sum_i X_i^2}{nb} - \frac{T^2}{nab} \qquad (5.25)$$

$$SS_b = \frac{\sum_j X_j^2}{na} - \frac{T^2}{nab} \qquad (5.26)$$

$$SS_{tr} = \frac{\sum_i \sum_j X_{ij}^2}{n} - \frac{T^2}{nab} \qquad (5.27)$$

Table 5.6 Basic ANOVA Matrix

Source of variation	Degrees of freedom	Sum of squares	Mean squares	F-value data	F-value
Factor A	$a - 1$	SS_a	$MS_a = SS_a/(a - 1)$	MS_a/MS_e	
Factor B	$b - 1$	SS_b	$MS_b = SS_b/(b - 1)$	MS_b/MS_e	
AB interaction	$(a - 1)(b - 1)$	SS_{ab}	$MS_{ab} = SS_{ab}/(a - 1)(b - 1)$	MS_{ab}/MS_e	
Error	$ab(n - 1)$	SS_e	$MS_e = SS_e/ab(n - 1)$	—	
Total	$(abn) - 1$	SS_t			
HO1: All $A_i = 0$		HO2: All $B_j = 0$			HO3: All $AB_{ij} = 0$

REFERENCES

American Society for Testing and Materials, 1976. *Presentation of Data and Control Chart Analysis* (STP-15D). Philadelphia, Pennsylvania.

Box, G. E. P., W. G. Hunter, and S. J. Hunter, 1978. *Statistics for Experimenters*, John Wiley and Sons, New York.

Deming, W. E., 1982. *Quality, Productivity, and Competitive Position*, MIT Press, Cambridge Massachusetts.

Edosomwan, J. A., 1986. Statistical process control in group technology production environment. Proceedings of SYNERGY '86, Conference on Functional Interfacing of Computer-Aided Manufacturing.

Grant, E. L., and R. S. Leavenworth, 1980. *Statistical Quality Control*, 5th ed. McGraw-Hill, New York.

Ham, I., 1976. Introduction to Group Technology. Technical Report MMR76-03. Society of Manufacturing Engineers, Dearborn, Michigan.

The IBM Quality Institute, 1985. *Process Control, Capability and Improvement*. The Quality Institute, Southbury, Connecticut, May.

Ishikawa, K., 1976. *Guide to Quality Control*, revised edition. Asian Productivity Organization, Tokyo.

Juran, J. M., F. M. Gryna, Jr., 1980. *Quality Planning and Analysis*, McGraw-Hill, New York.

Montgomery, D. C., 1976. *Design and Analysis of Experiments*, 2nd ed. John Wiley and Sons, New York.

Ott, E. R., 1975. *Process Quality Control*. McGraw-Hill, New York.

Rodriquez, R., and O. Adaniya, 1985. Group technology cell allocation. 1985 Annual International Industrial Engineering Conference Proceedings.

Roscoe, J. T., 1969. *Fundamental Research Statistics for Behavioral Sciences*, Holt, Rinehart and Winston, New York.

6

Productivity and Quality Planning

This chapter is concerned with the presentation of a conceptual framework and methodology for comprehensive productivity and quality planning in organizations. Basic forecasting techniques that will assist in productivity and quality planning are discussed. Technological forecasting is presented as a tool in the overall planning process.

6.1 COMPREHENSIVE PRODUCTIVITY PLANNING

Planning efforts exist in most organizations in a global sense. However, most of the plans are partial and based on extrapolation techniques. Most firms extend the annual operating plan into a 2-, 5-, or 10-year horizon. The most common source of projection is based on financial forecasts and business volume demand. In some situations, a linear extrapolation is made to project future spending and other resources that will be needed. The available productivity planning tools and methodology suffer from the following shortcomings.

1. They offer no logical framework for consistent planning within the organization.
2. They assume no significant changes in the regulatory environ-

ment. The risks associated with ventures are not completely monitored.

3. They assume no significant changes in economic factors, such as the inflation rate, resources, competition, and operating policies.
4. They offer no methodology for assessing the organization's strengths and weaknesses. There are few or no tools for a detailed overview of the firm or its industry and economic climate.
5. They offer no methodology for ongoing assessment of the threats and opportunities in the operating environment.
6. They offer no framework in which problems are easily identified and pinpointed to a base planning process. In some situations, objectives and goals arise by default and firms have to deal with inadequate planning programs through crisis management. This has a negative impact on total productivity and quality and the overall effectiveness of the organization.

According to Cotton (1976), "In productivity, planning is everything." Cotton recommends a three-step procedure for productivity planning:

1. Develop an effective planning process and structure in the organization.

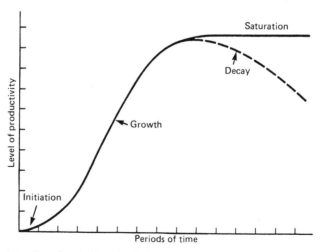

Figure 6.1   Productivity development curve. (*Source:* Cotton, 1976.)

Table 6.1   A Practical Eight-step Productivity Management Program Planning Process

---

Precondition or step: strategic and tactical planning awareness for the corporation and affected groups

Step 1: Internal strategic audit (looking within): What internal factors (strengths, weaknesses, conditions, trends, persons, programs, assumptions, etc.) should be considered during the decision, development, and potential implementation of our productivity management program?

Step 2: External strategic audit (looking around): What external factors (competitors, strengths, weaknesses, trends, conditions, organizations, assumptions, etc.) should be considered during the design, development, and potential implementation of our productivity management program?

Step 3: Planning premises, assumptions: importance—certainty grid.

Step 4: Strategic planning—2–5 year goals and objectives (desired outcomes) for the productivity program.

Step 5: Prioritization and consensus of performance objectives in key results areas relative to the productivity program.

Step 6: Identification, prioritization, and consensus for / of strategic, tactical and operational action programs: What programs, plans, resources, etc., will have to be budgeted for in both the short run (1 year) and longer run (2–5 years) in order for this program to succeed? Simultaneous operational focus (1 year) with a strategic (innovation, growth, continued development and support) focus (2–5 years). *Note*: Some programs will double in 1 year; others will take 2–5 years or more. Therefore, some subprograms are to be budgeted for in year 1 but are the basis for longer term payoff programs.

Step 7: Program planning and resource allocation.

Step 8: Program evaluation, review, maintenance.

Source: Sink, S. 1985. Strategic planning: A crucial step toward a successful productivity management program. *Industrial Engineering*, January, pp. 52–60. Reprinted with permission.

2. Prepare productivity goals and permeate the planning process with specific objectives based on these goals.
3. Establish productivity surveillance, assistance, and coordination in a manner tailored to the organization's need.

As shown in Figure 6.1, Cotton identified the various stages of productivity programs and projects. He pointed out that initiation, growth, saturation, and decay are common patterns that are typical of all productivity programs and projects. Sink (1985) presented an eight-step productivity management program planning process, shown in Table 6.1. He also presented six stages through which many productivity programs have evolved or will evolve.

Stage 0   The basics—refocus on fundamental management and engineering principles and practices

Stage 1   Organizational systems performance measurement—focus on performance measurement and evaluation systems at individual, group and organization levels

Stage 2   Integration of productivity planning with business planning—expand the scope and improve the effectiveness of strategic and operational planning systems

Stage 3   Participants in planning, problem-solving, design, and decision-making—design, develop, and experiment with participative management processes

Stage 4   Productivity measurement systems refinement—enhance the sophistication of performance and productivity measurement systems

Stage 5   Maintaining excellence—continued and ongoing maintenance and development, sponsor and promote appropriate levels of change and innovation

6.2   DEFINITION AND BENEFITS OF COMPREHENSIVE PRODUCTIVITY PLANNING

Edosomwan (1986) defined comprehensive productivity planning as a process by which all factors affecting an organization are considered in formulating its goals and objectives, assessing its capabilities and capacities, designing alternative courses of action for the purpose of achieving these goals and objectives, initiating necessary actions for their implementation, and evaluating the effectiveness of the plan. Comprehensive productivity planning provides the following benefits.

1. It minimizes resistance to change, reduces fear of the future, and stimulates creative thinking.
2. It permits the use of objectives to analyze and prepare realistic actions.
3. It serves as a formal means of coordination and integration in establishing a productivity program.
4. It sets the framework for identifying future productivity improvement possibilities and prepares for their possible adoption.
5. It creates a basis for effective supervision of necessary actions and improves decision-making by providing a better understanding of the future so that future decisions can be made more rapidly and more economically.
6. It provides a reference framework for measuring actions and initiating corrective measures.
7. It provides a basis for evaluating the operational feasibility of projects, new ventures, and policy changes.
8. It provides a basis for identifying the threats to and the strengths, weaknesses, and opportunities of the organization.
9. It provides a data base for auditing the effectiveness of staff programs and the restructuring of the organization to meet plan objectives.
10. Policies are refined and objectives set based on the diagnosis of actual events in the organization.

6.3 FEATURES OF COMPREHENSIVE PRODUCTIVITY PLANNING

The seven major features of comprehensive productivity planning are as follows.

1. Comprehensive productivity planning must be based on viable assumptions.
2. Comprehensive productivity planning must have stated achievable goals and objectives.
3. Creating awareness is an important factor in comprehensive productivity planning, especially for cooperation within an organiza-

tion. This can be done by explaining and defining goals and objectives to management, union, and employees.

4. Comprehensive productivity planning can have such dimensions as time (short range, medium range, and long range); organizational identification (such as certain parts of the organization—production planning, personal planning, and others); orientation (internal and external); and scope of the plan (all factors considered—comprehensive plan; few factors considered—partial plan).

5. Comprehensive productivity planning requires elements of flexibility to be built into the design, and it implies standards that become the basis for controlling the program.

6. Comprehensive productivity planning requires the input of forecasted variables since it deals with the future to make complex and interdependent decisions.

7. Comprehensive productivity planning requires suggestions from management, employees, and unions in designing alternative courses of action.

6.4 COMPREHENSIVE PRODUCTIVITY PLANNING AS A COMPONENT IN THE PRODUCTIVITY MANAGEMENT TRIANGLE

Comprehensive productivity planning is one of the components of the productivity management triangle (PMT) shown in Figure 6.2. The productivity management triangle encompasses an information system that provides input of information relevant to the planning and analysis process, performance and measurement evaluation and control process, and improvement monitoring process for the implementation of corrective actions and techniques that improve productivity. The productivity values and indices derived in the measurement phase are used for both short-term and long-term productivity planning. Forecasting techniques are used to determine productivity values and indices for future time periods.

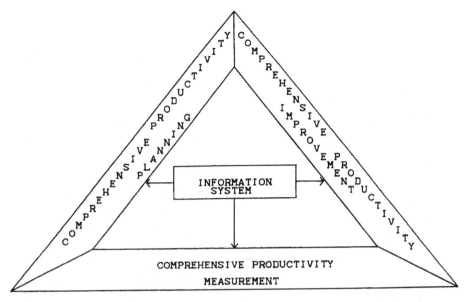

Figure 6.2 Components of the productivity management triangle. (*Source:* Edosomwan, 1986, p. 68. Reprinted with permission.)

6.5 COMPREHENSIVE PRODUCTIVITY PLANNING STAGES

Comprehensive productivity planning is not a one-time action but a continuous process consisting of four interrelated stages: productivity planning appraisal, strategic productivity planning, tactical productivity planning, and operational productivity planning. These four stages are the components of the comprehensive productivity planning cycle (CPPC) shown schematically in Figure 6.3.

For an organization starting a comprehensive productivity planning program for the first time, a productivity planning appraisal is the first stage in the comprehensive productivity planning cycle. Once the factors affecting the organization internally and externally have been assessed, a set of strategies is formulated. Based on these strategies,

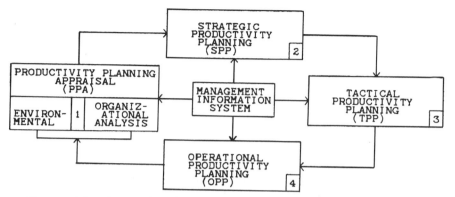

Figure 6.3 Comprehensive productivity planning cycle. (*Source*: Edosomwan, 1986, p. 68. Reprinted with permission.)

short-range objectives are formulated. Operational techniques are then used to implement both short- and long-range objectives. In order to assess the extent to which the objectives were successfully implemented, an organizational analysis is performed again. The cycle thus continues as long as the comprehensive productivity planning program exists in the organization.

6.6   IMPLEMENTATION METHODOLOGY

In order to implement a comprehensive productivity planning program, major factors affecting the organization in each stage of the comprehensive productivity planning cycle must be understood. Following this, the economic impact of the plan is assessed. In Figure 6.4, the implementation steps for CPPC are presented. Each step will now be described.

6.6.1   Establishment of a Comprehensive Productivity Program

In order to get started with a formal comprehensive productivity planning program, a planning council should be established. The council will develop basic implementation strategy, including training requirements, and focus on key areas of priority and structure that accommodate the four components of CPPC. The role of the council will

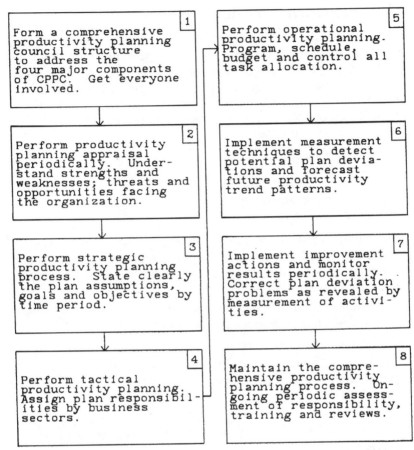

Figure 6.4   Implementation steps for a comprehensive productivity planning program.

also include providing technical leadership to the planning teams, facilitating total involvement of all persons concerned (management and employees), reviewing plan assumptions, goals, objectives, and accomplishments, and facilitating clear communication upward and downward within the organization. The planning council acts as the focal point for the organization on all planning issues. It is recommended that the council be comprised of members from all key functional areas of the business.

6.6.2 Productivity Planning Appraisal Process

The purpose of the productivity planning appraisal process is to understand threats and opportunities in the environment in which the organization operates and the strengths and weaknesses of the organization's internal operating characteristics. A productivity plan is then developed to address both the threats and the weaknesses. Action plans that will continue to maximize the opportunities and strengths of the organization are enforced.

Environmental Factors

In order to understand the threats and opportunities in the environment in which a company operates, the following environmental factors must be considered.

Economic trends: this includes the stage of the business cycle, inflationary or deflationary trends of prices, monetary policies, interest rates, tax rates, balance of payments, and surpluses or deficits relative to foreign trade

Government influence: including legislation on business operations by the federal, state, and local governments

Legal system: laws and regulations affecting business operations

Market conditions: changes in population, age distribution of the population, income distribution of the population, and product life cycles

Technological conditions: improved technology affecting raw materials, production method, production processes, and products or services

Social conditions: focuses on values and attitudes of customers and employees

Geographic conditions: includes site location, population, transportation, and availability of labor and resources

Organizational Factors

In order to identify the strengths and weaknesses of the company's internal operating characteristics, the following organizational factors must be considered: Management structures and philosophy, financial strength, communication flow pattern, production processes, technology utilized, research and development, resource requirements

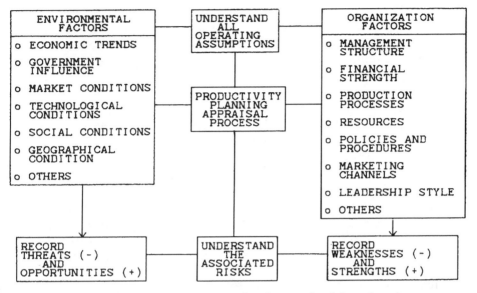

Figure 6.5 Factors and elements in the productivity planning appraisal process.

(human, materials, energy, and capital), organizational policies, leadership styles, management goals, and marketing and distribution channels. Based on the outcome of the environmental and organizational factor analysis, a productivity plan that addresses both the threats and the weaknesses is then put in place. Action plans that will continue to maximize the opportunities and strengths of the organization are enforced.

The factors involved in the productivity planning appraisal process are shown schematically in Figure 6.5.

6.6.3 Strategic Productivity Planning Process

Strategic productivity planning is concern with long-range goals and objectives, and it is broad. It refers to the process of setting the basic purpose and direction needed by an organization. Based on the appraisal performed in stage 1, the strategies and goals of the organization are formulated over the long run. To formulate goals and objectives, understanding corporate philosophy and operating strategies is necessary. This is useful for the evaluation of productivity improvement ideas. The corporate philosophy usually consists of the objectives, operating procedure, and environmental constraints. Objectives are interacting rather than independent; hence this is a complex and dynamic process.

Guidelines for Formulating Objectives

1. Productivity planning objectives must be based on viable assumptions. The knowledge obtained from the environmental and organizational analysis becomes crucial in formulating assumptions about such items as product demand, plant capacity, and resource requirements for production, sales, and services.

2. Productivity planning objectives must have achievable end results that are measurable. A single source of direction of the objectives reduces confusion within the organization. The partial, total-factor, and total productivity indices are good measures of an organization's performance and are recommended by Edosomwan (1985). Such measures have already been presented in Chapter 4.

3. Productivity planning objectives must be realistic, attainable, and based on identified known opportunity. The focus should be on setting objectives that are obtainable within the range of performance capability and resource availability.

4. Productivity planning objectives must be controllable through assigned accountability. The objectives must be reducible into various tasks that are assignable to work groups, departments, and individuals. Check points must be put in place to measure performance.

5. Productivity planning objectives must constitute three elements. Attribute, which specifies the dimensions along which the achievement of the objectives are to be measured; scale, which provides a yardstick for measurement at all levels; and goals, which specify value on the scale to be obtained within a specific time frame.

The elements involved in the strategic productivity planning process are shown schematically in Figure 6.6.

6.6.4   Tactical Productivity Planning Process

This stage of productivity planning is short range and concerns itself with the means of implementation. Tactical productivity plans support strategic plans and are relatively flexible. It is used to make decisions on such areas as production control, inventory control, resource re-

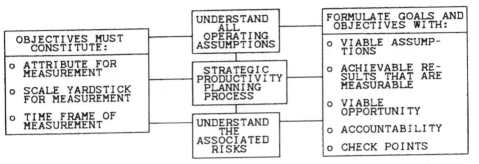

Figure 6.6   Elements in the strategic productivity planning process.

quirement planning, labor planning, information planning, and purchasing. Tactical productivity planning uses models in its planning process. The models are used in this stage to describe, explain, and predict the behavior of the system they represent. Based on specific organizational need, analytic models (linear programming, nonlinear programming, quadratic programming, dynamic programming, integer programming, inventory models, queuing models, Pert/CPM models, reliability models, and Markovian decision models) or simulation models (Monte Carlo technique, heuristic technique, and operational gaming technique) may be used as an aid in the planning process. In this step the productivity plans are carefully evaluated and redistributed by business sector. The assumptions and resources supporting the plans are streamlined to suit short-range business demand. The key elements involved in this process are shown in Figure 6.7.

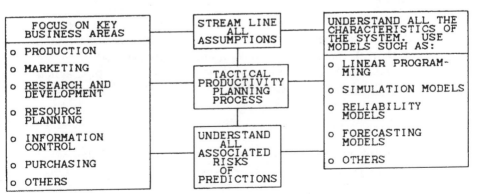

Figure 6.7   Elements in the tactical productivity planning process.

6.6.5   Operational Productivity Planning Process

Operational productivity planning is the last phase of the cycle. It is an operational, as opposed to a preparational phase. The other phases have all contributed toward specifying the design and laying the foundation for the development of the operating plan.

The purpose of operational productivity planning is to assign specific actions to specific groups with complete instructions. In order to generate reliable plans, the ability of each group must be recognized. The assignment process must take into account a group's access to needed information, needed data, qualification, and experience. The delegated actions require appropriate tracking means. The data and information processing must specify means for collecting data, processing, storing, and transmission. Provisions also have to be made for reporting unplanned events. However, it should be determined the event to be reported and the format in which to report it. Since operational productivity planning concerns itself with determining and delegating implementation actions and controlling them, the following functions are performed in this stage.

Programming

This is the function of determining the specific action steps necessary to attain the formulated goals and objectives. Designing alternative courses of action and choosing the best or optimum from among the alternatives is an effective programming method. One way of doing this is to test each action based on its feasibility and potential impact.

Scheduling

The scheduling function is to fit a timetable to the action steps selected. Scheduling assigns various action steps to the facilities required in such a way that all costs associated with the productivity program are minimized. Scheduling enables actions to be completed on time, action steps to be performed accurately, and resources to be utilized effectively.

Budgeting

The budgeting process provides a comprehensive financial plan based upon predetermined goals and objectives. The budgeting process also provides a means for assuring financial coordination by providing a mechanism for comparing revenues and expenses. The accuracy of the budget depends on a comprehensive management information system, an accurate accounting structure, and a clear understanding of the interdependencies among functions.

Controlling

Controlling the course of actions should be made an integral part of operational productivity planning. No plans will have a substantial impact on the program if they are not kept under proper surveillance. The controlling function uses a control system to achieve its objectives. This control system can be defined as a set of activities that maintains, in terms of predetermined goals, ongoing surveillance of the results of a process and attempts to correct the process when actual results are different from planned results. It can be viewed, mechanically, as consisting of three basic parts. A control sensor is used in reporting after identifying the actual state of the process under control. A control monitor is used for comparing the report (actual result) against the expected result. The output of this process should also show a variance between actual and expected results. The effective monitoring process must have a stated target in the dimension of the goal and a comparison framework. Finally, there must be a control analyzer that provides a means for analyzing deviations and taking corrective action aimed at the reduction in the magnitude of the variance. The elements involved in the operational productivity planning process are shown in Figure 6.8.

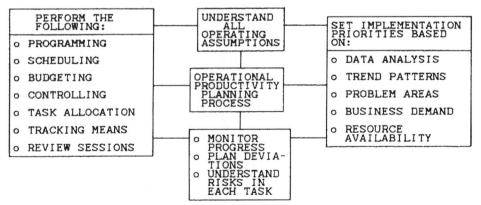

Figure 6.8   Elements in the operational productivity planning process.

6.6.6   Measurement Technique Implementation

In order to assess how well the plans were implemented, measurement techniques, such as total, total-factor, and partial productivities, should be implemented. Edosomwan (1985) developed and tested a productivity measurement model that is unique in this type of situation. The model is presented in Chapter 4. These measures should be used in conjunction with other types of financial measures, such as the balance sheet rate of return on investment and risk assessment based on cash flow. Record and take action on all potential problem areas.

6.6.7   Improvement Action Implementation

In this step, improvement actions that address all potential problems identified in Section 6.6.6 are implemented. Improvement action can vary from technical actions to people-oriented or customer-oriented actions. It may be task based or technology based. The key is to track the impact of all actions implemented and make modifications as necessary.

6.6.8   Project Maintenance

Comprehensive productivity planning is not a one-time action. It involves an ongoing assessment of the organization requirements associated with the components of the CPPC. Periodic review, training, competitive analysis, economic monitoring, and other tasks must be done continuously. Data collection and information systems should be implemented to ease the burden of tracking historical issues.

6.6.9   Setting Comprehensive Productivity Implementation Priorities

Since comprehensive productivity planning for the entire organization is time consuming, it may be necessary to direct efforts to areas in which benefits will be most apparent. Cost consideration is usually the basis for this priority. Problem areas and business demands are other factors to be considered. The productivity planning activities must be understood and supported by those charged with their implementation. Table 6.2 suggests how the comprehensive productivity planning stages could be assigned in a large organization.

Table 6.2   Allocation of Comprehensive Productivity Planning Tasks

Component of CPPC	Responsibility	Dynamic time duration for ongoing assessment
Stage 1: productivity planning appraisal	Usually achieved through task force; could be directed by site or by corporate headquarters	0–2 years, periodic
Stage 2: strategic productivity planning	Corporate headquarters	5–10 years, long range
Stage 3: tactical productivity planning	Division headquarters	2–5 years, medium range
Stage 4: operational productivity planning	Plants, departments, groups	0–2 years, short range

Source: Adapted from Edosomwan, 1986. Reprinted with permission.

6.7   COMMON PROBLEMS ENCOUNTERED DURING IMPLEMENTATION AND WAYS TO COUNTER THEM

Envisioning the future based on today's operating condition is often not an easy task. As a result, various forecasting techniques are used as the basis for predicting the future. Forecasted values for business volume demand and resources are likely to cause errors in any of the stages of the comprehensive productivity planning cycle. To overcome this problem, forecasted values should be supplemented by managerial judgment and experience with issues that require planning. Forecasting should be performed with diligence and with the best available skills and judgment. The best forecast is normally obtained through teamwork and good sources of data collection. Also, the method of forecasting has to be carefully selected. Erratic changes in corporate

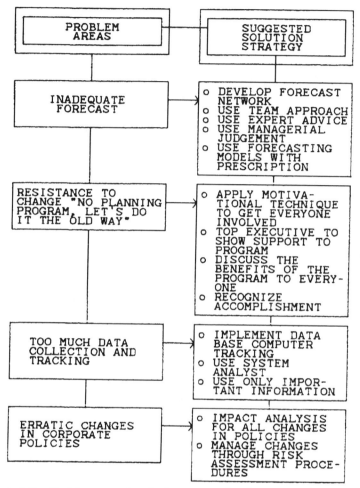

Figure 6.9   Problems and solution strategies in comprehensive productivity planning implementation

policies and objectives can have a negative impact on the productivity planning process. This problem can easily be corrected by top management. Changes in policy and objectives must be accomplished with an analysis of how they will affect the organization, the plan assumptions, and activities. Resistance to change is another common problem when trying to establish a comprehensive productivity planning program. A teamwork approach that involves all levels of management and employees should be adopted. This will eliminate the problem of the "It is not my plan" syndrome. Specific planning tasks should be assigned to specific people, and meetings should be held regularly to discuss and resolve problems of potential deviations from the plan. Productivity planning is not a one-time action. It is a continuous process that requires total organizational involvement in constantly analyzing business trend patterns and formulating and implementing new objectives that provide competitive operating advantages for the organization. The common problems usually encountered and solution strategies for handling them are presented in Figure 6.9.

6.8 RELATIONSHIP BETWEEN PRODUCTIVITY PLANNING AND MEASUREMENT

As presented in Chapter 1, comprehensive productivity planning is one of the components of the productivity and quality management triangle. The productivity values and indices derived in the measurement are used for both short-term and long-term productivity planning. Forecasting techniques are used to project productivity values and indices for future time periods. Regardless of the forecasting technique used, managerial judgment still plays a major role in the overall planning process.

6.9 BASIC FORECASTING TECHNIQUES FOR PRODUCTIVITY AND QUALITY PLANNING

The ongoing assessment of the productivity and quality planning program involves a dynamic time duration. Productivity and quality measures of current time periods should be projected for future time periods for continuous short- and long-range planning and detailed assessment of such projection on the overall business. Forecasting as a tool in the planning process provides the basis for accomplishing such objectives. Some of the forecasting techniques will now be described.

6.9.1 Last Value Forecasting Technique (LVFT)

This technique projects the productivity value or indices and the yields of a particular process or task based on the last value of previous time periods. Let

$\hat{P}_t$ = productivity value or indices forecast for a given task or process in period t

Q_t = quality yield or characteristics forecast for a given task or process in period t

P_{t-1} = actual productivity value or indices for a given task or process in the last period

Q_{t-1} = actual quality yield or characteristics for a given task or process for last period

The LVFT proposes that

$$\hat{P}_t = P_{t-1} \tag{6.1}$$

$$Q_t = Q_{t-1} \tag{6.2}$$

The LVFT has the disadvantage of being based upon a sample size of 1. The estimate often has wide variance and can be imprecise. The LVFT is useful only in situations in which the process is changing very rapidly so that any value before time t is almost irrelevant. The technique is also useful if the conditional distribution of the process concerned has a small variance.

6.9.2   Arithmetic-Average Forecasting Technique (AAFT)

This technique uses historical data to project the productivity values or indices and quality yields for a future time period

$$\hat{P}_t = \sum_{i=1}^{N} \frac{P_t}{N} \tag{6.3}$$

$$\hat{Q}_t = \sum_{i=1}^{N} \frac{Q_t}{N} \tag{6.4}$$

where:

$\hat{P}_t$ = productivity values or indices forecast for a given task or process for period t

P_t = historical data on productivity values and indices for given task or process

$\hat{Q}_t$ = quality yields for a given task or process for period t

N = number of study period

$t = 1, 2, 3, \ldots, K$

Q_t = historical data on quality yields for a given task or process

The AAFT relies on large masses of data. The technique is useful in situations in which the data are stationary and randomly distributed. One major disadvantage of this technique is that the computations and tracking required for the data can be cumbersome. Also, when large data included in the forecast involve process occasional shifts periods, the forecast values may be misleading.

6.9.3 Moving-Average Forecasting Technique (MAFT)

MAFT generates the next period productivity values or indices and quality yields for a given task or process by averaging the actual values for the last n periods.

$$\hat{P}_t = \frac{1}{N} \sum_{i=1}^{N} P_{t-1} \qquad (6.5)$$

$$\hat{Q}_t = \frac{1}{N} \sum_{i=1}^{N} Q_{t-1} \qquad (6.6)$$

where:

P_{t-1} = last period productivity value or indices for a given task or process

Q_{t-1} = last period quality yields for a given task or process

N = specified moving average period, for example, 2 months, 3 months

The moving-average forecasting technique uses relevant historical data. The estimator has one major advantage. It uses recent history and represents multiple observations. Although the technique is good for point analysis and long-range analysis, it responds to trends with delay and does not compensate for seasonal changes.

6.9.4 Exponential Smoothing Forecasting Technique (ESFT)

ESFT places more emphasis or recent data. The weight given to past data decreases geometrically with the increasing age of the data. This technique estimates the average productivity and quality values by adding to the last average forecast a fraction α of the difference between actual values and average forecast. The ESFT is good for treating autocorrelated data.

Let:

P_t = average new forecasted productivity value or indices for a given task or process

N = number of periods

P_{t-1} = actual productivity value indices for a given task or process

P_{t-1} = average productivity forecast values and indices

= exponential smoothing constant $0 \leqslant \alpha \leqslant 1$

E_{t-1} = error in previous forecast

New productivity values and indices = old forecast + actual forecast − old forecast

$$P_t = P_{t-1} + (P_{t-1} - P_{t-1}) \tag{6.7}$$

$$E_{t-1} = P_{t-1} - P_{t-1} \tag{6.8}$$

$$= \frac{2}{n + 1} \tag{6.9}$$

Choosing the right α may pose a problem. If α is too small, it responds to change slowly. The most commonly used values of α are 0.2 and 0.3. When past history is available, a simulation technique may be used to select the appropriate α.

6.9.5 Choosing the Best Forecasting Technique

The mean absolute deviation (MAD) is one method of selecting the best forecasting technique. The mean absolute deviation is determined by dividing the sum of absolute deviations by the number of observations:

$$MAD = \sum_{i=1}^{N} \frac{(/P - P_i/)^2}{N} \tag{6.10}$$

where:

P_i = actual productivity values or indices

P = forecasted productivity values or indices

N = number of observations

Each forecasting technique is applied to the historical data, and the technique with the smallest MAD is selected. The forecast error Fe can also be used to choose the technique:

$$Fe = \sum_{i=1}^{N} \frac{P - P_i}{N} \tag{6.11}$$

6.9.6 Factors Affecting the Choice of Forecasting Technique

The following factors can determine the applicability of forecasting techniques to a given data set:

1. Forecasting accuracy based on minimum data requirements
2. Length of time for which projections are made: short, medium, and long range
3. Patterns of variation present in the historical data

6.9.7 Steps in the Forecasting Process

A six-step approach that can be used in the forecasting process is shown schematically in Figure 6.10.

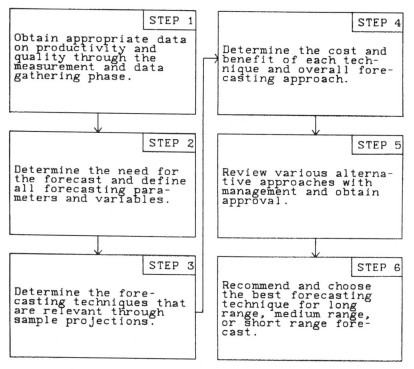

STEP 1	STEP 4
Obtain appropriate data on productivity and quality through the measurement and data gathering phase.	Determine the cost and benefit of each technique and overall forecasting approach.

STEP 2	STEP 5
Determine the need for the forecast and define all forecasting parameters and variables.	Review various alternative approaches with management and obtain approval.

STEP 3	STEP 6
Determine the forecasting techniques that are relevant through sample projections.	Recommend and choose the best forecasting technique for long range, medium range, or short range forecast.

Figure 6.10 Steps in the forecasting process.

6.10   QUALITY PLANNING

Comprehensive quality planning involves a process of assessing all factors that could affect product or service quality throughout the specified life cycle. The four components of the productivity planning discussed in Section 4.5 also apply to quality characteristics. The only major difference is that several additional aspects must be managed.

1. A program for product and service reliability planning must be instituted. Such a program provides appropriate planning for reliability prediction, design review, reliability testing, failure analysis, defect isolation, stress analysis, identification of critical parts as part of the methodology for reporting failure, and overall reliability goals for long-, medium-, and short-range planning.

2. A program for quality in planning the business process. Current problems of quality will change over time. There must be strategies for reducing process defect levels in the future, projections of process yields, and management tools to plan for their re-evaluation. Quality must be obtained at the source of production or service at a cheaper cost through appropriate sets of goals or objectives implementable over time.

3. Vendor management planning. The last two decades has seen emphasis on using vendors as a focal point for product suppliers and manufacturing and service operational units. The planning aspects in vendor quality management must encompass methodology, processes, and procedures to improve vendor-related defects, with the ultimate goal of zero defects in all products delivered on time to the required source. When strong reliance on vendors calls for specifications, there must be a strategy for future product quality based on how the specification might change over time. Vendor selection strategy most also encompass a list of defects and the quality expected from various vendors.

4. Planning for customer quality requirements. The goal should be a defect-free product to the customer at all times. A planning approach to ensure this must involve product serviceability, maintainability patterns, a built-in trigger backup for products, and ongoing assessment of the customer environment. The key is to understand how the customer's requirements will change over time. A strategy for quality that enables design, processes, production, and services to meet such a demand are then planned for the future. All new products or services must be better than those they are replacing. Customer quality re-

quirements can be assessed through interviews of the sales force or by questionnaires specifically designed for product and quality characteristics. The results of such questionnaires should be referred to a common period to enable realistic planning for quality variables.

5. Planning based on technical vitality and skills: This may be one of the most important aspects that determine the quality of the product or service at any given point in time. As a product or service vendored changes over time, the appropriate skills required to produce or offer a good product or service must be assessed and planned for. The most reliable information needed for skills planning is the processes and phases of the product or services. The key planning question that must be answered is: What skills are required for each phase, each process, and each operational unit of the future product or service?

6. Quality policies and objectives. At all levels of management and function, there should be policies that provide guidelines for planning the overall quality program and for specifying key quality attributes and characteristics that may change over time. At the organizational level, a policy statement must be made and goals for the future must be clearly stated and defined with task responsibility as it relates to the planning aspect assigned to a specific group with complete instructions. Quality objectives must encompass strategies that provide a basis for leadership through product performance, eliminate a poor image with customers and vendors, reduce reliance on inspection and rework reduction, require less field and service support, lessen complaints from customers, and reduce the cost of the product warranty. At the operational level, it is recommended that each department or operational unit formulate its quality goals and objectives that compliment the overall organizational goals and objectives.

7. Quality information system. In order to plan and make realistic projections on quality issues, a quality planning information system must be instituted. Such a system will track and provide the basis for the analysis of audit results, design evaluations for quality, process data, inspection data, market research information on quality, purchase parts and materials characteristics, specifications for processes, and tools, field performance data, and product design test data. Computer-based quality information is recommended. Manual tracking has a major disadvantage of excessive manual computations, tracking, and control. However, the computer-based systems, if implemented, must also be able to provide information on real time with minimum system downtime. The quality of the computer software can

Table 6.3 Matrix for Executive Reports on Quality[a]

Control subject	Typical unit of measure	Usual departmental source of data	Typical standard	Typical format	Typical frequency
Negative customer reactions					
Complaints	Number of complaints per 1000 units; per $1000 of sales	Field service	Historical	Narrative	M
Returns	Value of returns per $1000 of sales	Accounting	Historical, market	Narrative	M
Service calls	Number of service calls per 1000 units under warranty; cost of service calls per 1000 units under warranty	Field service	Historical, market	Tabulation	Q
Guarantee charges	Dollars per 1000 units under warranty	Accounting	Historical, market	Tabulation	Q
Field performance					
Product reliability	Failure rate; mean time between failures	Field service	Engineered, historical, market	Charts	M
Spare-parts sales	Dollars of sales	Accounting	Historical	Tabulation	A
Product conformance on inspection, test	Defects per unit; process average for specific qualities	Inspection, test	Historical, engineered	Charts	M
Outgoing quality based on product audit	Demerits per unit	Quality assurance	Historical, market	Charts	M

Subject	Unit of measure	Source	Comparison base	Presentation form	Frequency[a]
Vendor quality performance	Dollars of cost per dollars of purchases	Accounting, quality control	Historical	Tabulation	Q
Quality costs: appraisal, failure, prevention	Dollars per hour of direct labor; per dollar of direct labor, shop cost, processing costs, sales; per unit of product, equivalent product	Accounting, quality control	Historical, budget	Tabulation, charts	Q
Surveys; audits other than product audit	Various	Quality assurance	Plan	Narrative	S
Opportunities	Return on investment, other	Quality control	—	Narrative, tabulation, charts	Q
Customer relations on quality (other than alarm signals)	Various	Marketing, field service, quality control	—	Narrative	Q
Results of quality improvement programs	Dollars, return on investment	Quality control	Plan	Tabulation, narrative, charts	Q

[a]Frequency code: M = monthly; Q = quarterly; A = annual; S = special.

Source: Adapted from Juran, J. M., and Gryna, F. M., Jr., Quality Planning and Analysis, 2nd ed. McGraw-Hill, New York, pp. 520–521. Reprinted with permission.

have significant impact on the overall quality of the data base management. Examples of some types of reports on quality are shown in Table 6.3.

6.11   TECHNOLOGICAL FORECASTING: A KEY INGREDIENT IN THE OVERALL PLANNING PROCESS

In the era of technological explosion, productivity and quality within a given time period in any organization will be driven largely by the types of technology involved. It is important for the organization to have a formalized program for technological forecasting. Such a program can provide the following benefits:

1. Help to establish the timing of new technology and maximize gain from events that are the result of action taken by the corporation
2. Set quantitative performance standards for new products, processes, and materials
3. Assist in the planning of research programs: the amount and direction, the scientific skills needed, and resource utilization
4. Guide engineering programs toward the use of new technology and the adjustment of new technical demand
5. Help in new product development and current product improvement
6. Help to identify major opportunities and threats in the technological environment and their social impact on employment, skills, and educational needs, for example
7. Help to identify the economic potential and impact of technological progress and guide technological planning and its contribution to long-range planning
8. Help to determine leadership in industry and increase the future market share and profit of the corporation

6.11.1   Definition of Technological Forecasting

Technological forecasting is a tool used for the prediction and estimation of the feasible or desirable parameters in future technologies. In high-technology industry, technological forecasting is seen as a prediction of what might happen given certain assumptions and objectives.

6.11.2 Operational Definition of Technological Forecasting by Levels in High-Technology Industry

Technological forecasting can be defined according to the operational level:

Policy planning. The clarification of scientific technological elements determining the future boundary conditions for corporate development.

Strategic planning. The recognition and comparative evaluation of alternative technological options.

Operational or tactical planning. The probabilistic assessment of future technology.

Marketing and corporate profit planning. The clarification of scientific technology needed to expand the market share and compete in the world market.

6.11.3 Technological Forecasting Methodology in High-Technology Industry

It is important to point out that, owing to the great competition that exists within high-technology industry, a logical framework is needed to perform adequate forecasts. Technological forecasting methodology is presented schematically in Figure 6.11. Each component of the framework will now be described.

Where Technological Forecasting Begins

Technological forecasting begins with a strategy meeting of the management team. The purpose of the strategy meeting is to reach agreement and understanding of the fundamentals of the development of the various subforecasts, to assign responsibilities, and to set a schedule. The fundamentals include

1. A review of total corporate long-term and short-term objectives
2. An assessment of the present status and a relative forecast of the future technology needed for the total business environment
3. A statement of the objectives and motivation for the technological forecast period within the constraints and opportunities recognized in the total business environment

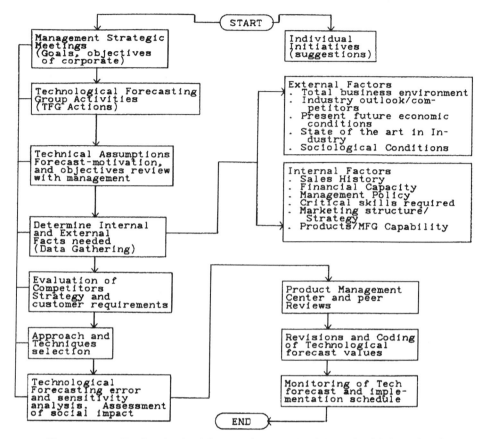

Figure 6.11   Technological forecasting methodology in high-technology industry.

Technological forecasting may also begin with individuals or group intuition. This usually occurs after on-the-job training in individuals with self-initiative and curiosity about what may happen in the future.

Technological Forecasting Group (TFG) Activities

Most major corporations in high-technology industry have dedicated people who have been given the responsibility for technological forecasting. The technological forecasting group is made up of various experts with interdisciplinary backgrounds. The TFG ensures that all technical assumptions and motivations for the technology are clearly defined; they determine the internal and external facts needed, evaluate competitor strategy and customer requirements, select approaches and techniques to perform the forecast, assess the social impact of the forecast, code variables and parameters, and monitor the technological forecast and implementation schedule.

Important Aspects of Technological Forecasting

1. Sales history
2. Management strategic plan and policy
3. Marketing structure and strategy
4. Current product classification and diversity
5. Manufacturing capability
6. Financial capacity
7. Critical skills required

Other Factors Influencing Technological Forecasting

1. Total business environment
2. Some environmental influences on business
3. Total industry outlook
4. Appropriate facts about competitors and their products and plans
5. Outside expert recognition

Some of outside influences on business were shown schematically in Figure 6.11.

Evaluation of Competitor Strategy, Customer Requirements,
and the State of the Art in Specific Areas of Interest

Some of the specific reasons for evaluating competitor strategy, customer requirements, and the state of the art in areas of interest are as follows.

1. What is available in high-technology industry, and what are the areas of interest?
2. What are the competitors doing?
3. What products satisfy the customer of today, and what product will satisfy the customer of tomorrow?
4. Is there consistency in manufacturing capability, market structure, and labor skills, for example?

5. What are the latest measuring techniques (scientific papers and research efforts)?

Technological Forecast Timing

In high-technology industry the range of the technological forecast period is usually dictated by the purpose of the forecast. The technological forecast horizon is required to cover the cumulative lead time required for executing the plans. The technological forecast can be classified as

Short-term technological forecast, < 2 years

Intermediate technological forecast, 3–4 years

Long-term technological forecast, 5–20 years

The frequency of technological forecasting and review is usually in intervals that are no longer than the length of the forecast period; for example, annual forecasts are updated quarterly, long-term forecast annually.

Means, Approach, and Techniques of Measurement

The two main approaches usually used for technological forecasting are (1) the exploratory approach (push approach): project the area of technological forecasting to the future; and (2) normative approach (pull approach): start with goals and objectives, then identify the end item or result.

Some Techniques for Technological Forecasting

In high-technology industry, statistical techniques, operation research techniques, and reliability analysis are heavily used in technological forecasting.

Single-trend extrapolation: This technique involves extrapolating the trend of certain technological parameters; for example, the memory capacity of a personal computer, maximum aircraft speed, and operating energy of particle accelerators are plotted over long periods as a function of time. It is then assumed that the nature of the progress experienced in the past will continue to occur in the future following the same trend.

Growth analogy. Growth analogy techniques assume that biologic growth provides a useful analogy for many technologies. For example,

Lenz (1968) applied the biologic growth analogy to a projection of trends in the maximum speed of military aircraft.

Regression and correlation. This is a statistical technique used in technological forecasting to forecast the values of certain unknown parameters given historical trends. The two most popular regression methods are simple linear regression and double linear regression.

Substitution. In the substitution forecast, instead of measuring the increase in performance occurring in technology, the rate at which one technology is substituted for another in general usage is measured.

Fitted curves. This technique involves the extrapolation of time-series-related trends. The fitted curves are used as an aid to forecast what could happen in the future.

Delphi. The Delphi technique is used to obtain a consensus of experts. This technique is designed to systematically combine individual judgments and thus obtain a reasoned consensus about the technological forecasting input needed. Questionnaires are used in most cases to obtain opinions.

Scenarios. Scenarios can either be a direct extrapolation of the present conditions or variations formed by adding new conditions to the present environment. The three methods of scenario are the consensus techniques, the cross-impact matrix, and the alteration-through synopsis. The cross-impact matrix, which permits the orderly investigation of the effects of potential interactions among items in a forecasted set of occurrences, is the most widely used.

Expert Evaluation of Technological Forecasts

Expert evaluation is a sanity check on the technological forecast group output. The procedure ensures

1. Consistency of technological forecasting objectives with corporate strategic direction: Why is this forecast needed?
2. Appropriate technological forecasting assumptions and intelligently interpreted facts
3. Appropriate information and data and intelligently interpreted facts
4. Validity of the approach and the techniques used
5. Comprehensibility of study
6. Assessment of social impact
7. Realizing the limitations of the technological forecast
8. Monitoring tools and techniques

9. Implementation schedule and recognition of pitfalls
10. Validity of the time horizon forecast

*Coding of Technological Forecast Variables, Parameters,
and Values*

One of the most important tasks of the technological forecast group is the coding of forecast variables, parameters, and values. This is done to

1. Prevent potential competitors from obtaining the technological forecast information, avoiding a loss of the market share or product leadership, for example
2. Prevent unqualified practitioners from misusing the technological forecast information, especially in such areas as defense systems and intelligence monitoring
3. Protect the product announcement schedule
4. Minimize the loss from internal and external data sales crime

Monitoring the Technological Forecast

Technological forecast measuring tools and the units of measure should be selected or designed to fit the subject, the product, the environment, the objective, and the user. Some examples of practical measuring tools for monitoring the status of technological forecast errors in high technology industry are (1) the coefficient of variation = standard deviation divided by the average projected forecast value (demand) and the standard error = standard deviation divided by 5N (number of observations); (2) volume index graph, line-average graph, and moving-average graph; (3) comparison statements; and (4) the company profit plan.

*Problems and Pitfalls in Technological Forecasting in High-
Technology Industry*

1. Inappropriate environmental study (gathering data on competitors and politics)
2. Lack of adequate data on industry averages in many technical specification areas
3. Distortion of the forecast by strong desire over specification or reliability information, for example
4. Poor technological forecast monitoring tools and techniques

5. Lack of total involvement from the management team and technological forecasting group members
6. Lack of documentation of the technological forecast

Keys to the Successful Use of the Technological Forecast in High-Technology Industry

1. Understanding the basic assumptions of technological forecast and their limitations and preparation to be flexible enough to manage the forecast successfully within an agreed and pre-planned tolerance range of forecast error
2. Timely measurements of technological forecast errors
3. Good management interpretation and understanding of differences between the technological forecasts and actual values
4. Management response to the "vital signs": timely recognition and understanding of developing problems and adjusting forecasted values and parameters to suit prevailing conditions
5. Total involvement of all the "players"

Summary of Key Issues in Technological Forecasting

Technological forecasting is important because it defines our best estimate of what could happen, the opportunities and restraints of our future. Usually all the subsequent planning of production, materials, and capacity will be developed with the scope and parameters of a corporate technological forecast.

We must identify information about the future that will be essential to technological forecasting:

Calculate any essential data that may be done with precision.

Forecast the balance of essential information that is required but cannot be calculated.

Technological forecasting requires the involvement and commitment of every member of the forecasting group and management team.

All assumptions on which technological forecasts are based should be defined and stated, understood and, all agreed to, by all members of the forecasting group and management team.

The risk-reward relationships involved in hedging against the margin of technological forecast error should be discussed openly, to avoid conflicts and to build credibility.

Good technological forecasting is accomplished by progressively gathering appropriate essential facts; interpreting the facts intelligently; modifying statistics with management judgment; monitoring the technological forecast error; and progressively updating and refining the forecasts.

Understanding and using the technological forecast is even more important management work than making the forecast. Decision makers are concerned with validity, credibility, comprehensibility, and accuracy.

Technological forecast accuracy depends upon many factors, including the qualitative and quantitative techniques utilized, the source of the data, forecaster know-how, a clear understanding of the business environment, and knowledge of the state of the art in the field.

Attention should be paid to significant differences between the technological forecast and the actual situation. These are usually early warning signals calling for corrective action.

Technological forecasting is an input to the planning process; it is not a plan in itself. A technological forecast tries to predict what may be technologically.

REFERENCES

Abell, D. F., 1980. *Defining the Business*. Prentice-Hall, Englewood Cliffs, New Jersey.

Ackoff, R. L., 1981. *Creating the Corporate Future*. John Wiley and Sons, New York.

Andrews, K. R., 1971. *The Concept of Corporate Strategy*. Dow Jones/Irwin, Homewood, Illinois.

Ansoff, H. I., 1965. *Corporate Strategy: An Analytic Approach to Business Policy for Growth and Expansion*. McGraw-Hill, New York.

Blanchard, K., and S. Johnson, 1982. *The One Minute Manager*. William Morrow, New York.

Bradley, J. W., and D. H. Korn, 1900. *Acquisition and Corporate Development, A Contemporary Perspective for the Manager*. Arthur D. Little, Lexington, Massachusetts.

Bright, J. R., 1968. *Technological Forecasting for Industry and Government*, Prentice-Hall, Englewood Cliffs, New Jersey.

Cetron, M. J., 1969. *Technological Forecasting, A Practical Approach*. Technology Forecasting Institute, New York.

Cetron, M. J., and A. L. Weiser, 1968. Technological change, technological forecasting and planning R&D—a view from the R&D manager's desk. *George Washington Law Review*, Technology Assessment and the Law, Vol. 36, No. 5, July, pp. 1090, 1091.

Cotton, F., 1976. In productivity, planning is everything. *Industrial Engineering*, November.

Edosomwan, J. A., 1985. A task-oriented total productivity measurement model for electronic printed circuit board assembly. International Conference on Electronics Assembly Proceedings, Santa Clara, California, October 7–9.

Edosomwan, J. A. 1985. A nine-step approach for long range productivity planning. Working paper, ASEMM.

Edosomwan, J. A., 1986. A conceptual framework for productivity planning. *Industrial Engineering*, January.

Ewing, D. W., ed., 1964. *Long-Range Planning for Management*. Harper and Row, New York.

Harrigan, K. R., 1900. *Strategies for Declining Businesses*. Lexington Books, D.C. Heath and Company, Lexington, Massachusetts.

Helmer, O., 1969. Analysis of the future: The Delphia method. In *Technological Forecasting for Industry and Government*, J. R. Bright, ed. Prentice-Hall, Englewood Cliffs, New Jersey. Dalkey, N. C., The Delphi method: An experimental study of group opinion. The RAND Corporation, RM-5888 PR, June, Santa Monica, California.

Henderson, B. D., 1982. *Henderson on Corporate Strategy*. Mentor Books, New York.

Jones, H., 1974. *Preparing Company Plans: A Workbook for Effective Corporate Planning*. John Wiley and Sons, New York.

Lenz, R. C., Jr., 1966. Technological forecasting. Paper presented at the U.S. Air Force Symposium on Long-range Forecasting and Planning, Colorado Springs, Colorado, August, pp. 155–157.

Lenz, R. C., Jr., 1968. Forecast of exploding technologies by trend extrapolation. In *Technological Forecasting for Industry and Government*, ed. by J. R. Bright, ed. Prentice-Hall, Englewood Cliffs, New Jersey, pp. 65–69.

Lorange, P., 1980. *Corporate Planning: An Executive Viewpoint*. Prentice-Hall, Englewood Cliffs, New Jersey.

Ohmae, K., 1982. *The Mind of the Strategist: The Art of Japanese Business*. McGraw-Hill, New York.

Ouchi, W. G., 1981. *Theory Z: How American Business Can Meet the Japanese Challenge*. Avon Books, New York.

Pearce, and Robinson, 1982. *Formulation and Implementation of Competitive Strategy*. Richard Irwin, Homewood, Illinois.

Porter, M., 1980. *Competitive Strategy: Techniques for Analyzing Industries and Competitors*. Free Press, New York.

Quinns, J. B., 1967. Technological forecasting. *Harvard Business Review*, Vol. 45, No. 2.

Quinn, J. B., 1980. *Strategies for Change: Logical Incrementalism.* Richard Irwin, Homewood, Illinois.

Rothschild, W. E., 1979. *Strategic Alternatives: Selection, Development and Implementation.* Amacom, New York.

Sink, S. 1985. Strategic planning: A crucial step toward a successful productivity management program. *Industrial Engineering*, January, pp. 52–60.

7

Quality Error Removal
Improvement Technique

This chapter presents a step-by-step methodology for implementing the quality error removal (QER) technique in the work environment. The QER technique provides guidelines for work groups in resolving problems at the source of production or service. Two case studies are presented. The first case study assesses the impact of the introduction and use of the QER technique on total productivity in an assembly task. The second case study demonstrates the application of the QER in a bank clerical task.

7.1 QUALITY CIRCLES

In the early 1960s, the Japanese started the quality circle (QC) concept, based on the motivational techniques developed by American researchers and practitioners. In the late 1970s, the quality circle concept began to receive unprecedented worldwide attention. This was primarily due to Japanese success in producing high-quality products at high levels of productivity using QC techniques. In recent years, many companies and organizations worldwide have adopted the QC technique. There are also several publications on the subject on QC, such as Dewar (1980), Robinson (1982), Doyle and Straus (1978), IBM (1982), and Edosomwan (1985a, b). In addition to allowing workers the freedom to resolve problems based on their own judgment, the QC concept facilitates teamwork. The concept has been used widely to improve productivity and quality in many organizations. The quality error removal technique developed by Edosomwan (1983) is a unique type of quality circle concept that focuses on resolving problems at the source of production or service.

7.2 BENEFITS OF THE QUALITY ERROR REMOVAL TECHNIQUE

When implemented in the work environment, the QER technique provides the following benefits.

1. The QER technique improves quality and productivity at the source of production or service.
2. It provides the basis for dividing large work contents into small achievable tasks that are manageable by work groups. Organizational goals are clearly simplified into small tasks at the operational level. Work groups assume total responsibility for their quality and productivity and feel an integral part of the goal-setting process and the end results.
3. The technique improves employee morale and job satisfaction. It provides the basis for employees to assume total control at the source of production. Problems are solved through a brainstorming approach at a much faster pace. Sources of help are readily available to work group team members.
4. The QER technique improves communication channels at all levels of the organization. Management is viewed as part of the team effort to remove the obstacles that affect the quality of production and services.
5. The greater sense of control provided through the QER technique improves employee commitment, promotes a sense of dedication, yields fewer turnovers and a lower absenteeism rate, and provides good feelings about job security.
6. The improved quality and productivity obtained through the QER technique translates to more profit for the organization through product or service cost reduction and lower prices paid by customers for goods and services offered.

7.3 QUALITY ERROR REMOVAL TECHNIQUE DEFINED

The quality error removal technique provides a framework, principles, and guidelines to a group of employees who voluntarily work together to select and solve key problem(s) affecting an organization's work unit or task. Organizational goals are broken down into small tasks at the operational level, and continuous effort is applied to improve productivity and quality within each work unit.

7.4 QER TASK BOUNDARIES

The task boundaries involved with the use of the QER technique are specified in Figure 7.1.

7.5 QER KEY PRINCIPLES AND GUIDELINES

The following are the key principles and guidelines used by the quality circle team.

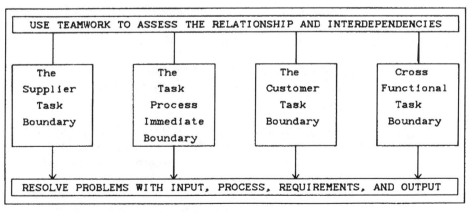

Figure 7.1   QER task boundaries.

7.5.1 Teach Basic Job Skills and Cross-Train for Multiple Skills

Both employees and management will be convinced that productivity and quality improvement is possible in an organization and within each task unit, once everyone involved understands how the job should be performed to deliver a good product. Everyone also understands the nature of the product or services being delivered. Two types of training are encouraged: formal training through a company educational program or an outside institution, and training by peers.

7.5.2 Encourage Decision Making and Problem Solving at the Source of Production or Services

At the operational level, the QER teams select their own problems on which to focus and decide how to attack the bottleneck issues that are preventing productivity and quality improvement. The team identifies problems over which there is control. All members of the team freely contribute ideas and reach a consensus on the best possible solution. The brainstorming technique is highly encouraged. The following steps are followed to understand the cause and effect of a key problem.

1. Precisely define the problem.
2. Draw a flow diagram, and label all categories.
3. Brainstorm to identify the root cause of the problem.
4. Use pareto diagram to select a problem on which to focus.
5. Obtain consensus on a solution.
6. Implement the solution and follow-up on issues.

7.5.3 Encourage Productivity and Quality Improvement at the Source of Production or Service

Each product or service inspection is done at the source. Inspection is on the line at the task level. Members of the QER team perform inspection of their own work. Quality and productivity errors are resolved by joint effort among the team. Teams have checkpoints to review all outgoing finished products. Any member of the team can call for group help to reach a consensus on key problems quickly. Feedback mechanisms between each QC team member are immediate. Team members are viewed as teachers and error-correction masters.

7.5.4 Encourage the Following QC Team Characteristics

All the members of the team clearly understand and agree on common goals. A forum exists for the sharing of ideas; there is a high level of communication among team members, and everyone participates with enthusiasm. Supervisors or managers support the team to resolve problems when called upon. Reward and recognition are shared among team members.

7.5.5 Clear Definition of Roles and Responsibilities of the QC Team

The definition of the roles and responsibilities are as follows:

Teacher: the person who trains team members for the skills needed to perform the task. At one stage or another, everyone on the QC team is a teacher.

Team leader: the person who leads the team through the process of resolving a key problem. The team leader is chosen by voluntary means. Each member of the QC team picks up leadership on a rotational basis.

Team members: a group of people who participate in the problem-solving and resolution process. The team meets regularly at specified times and a specified location to provide a consensus that resolves problems related to their tasks.

Facilitator: a team member who records all the team's progress, keeps records of the agenda, and assists the team leader in accomplishing the stated objectives and goals of the QC team.

Team correction master: any member of the team who identifies a quality error in the output produced by other team members. Such a member ensures that the error is corrected at the source of production or service. All team members participate in a random and revolving inspection process to help detect quality errors.

Team presentations: presentations to management on team accomplishments are rotated among team members. The presenter introduces team members at the start of the presentation. The agenda and the presentation package are jointly prepared by all members of the team.

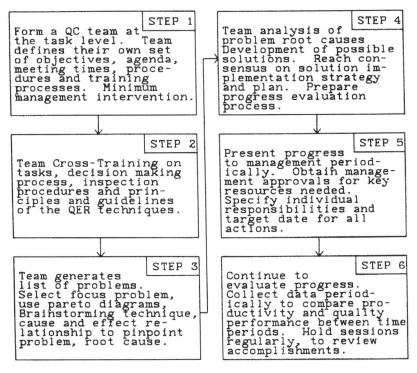

Figure 7.2   Implementation steps for the QER technique.

7.6   IMPLEMENTATION METHODOLOGY

The six steps shown in Figure 7.2 are suggested for the implementation of the QER technique.

7.7   CASE STUDY: IMPLEMENTATION OF THE QER TECHNIQUE IN AN ASSEMBLY TASK

The quality error removal technique was implemented at a manufacturing plant that produces printed circuit boards to customer order. Two groups of five subjects each were studied. The groups initially performed the printed circuit board assembly tasks without any formal training for the QER technique. After a period of 5 weeks, the subjects were trained on how to use the QER technique. The assembly process of printed circuit boards involved the normal insertion of components, such as transistors, diodes, modules, resistors, and capacitors, into an empty printed circuit board (PCB).

The work area was in an approved electrostatic discharge area that consisted of a grounded workbench, chair, tools, and a light box. The manual assembly process started with the verification of paperwork and parts. The component placement list (CPL), pick list, routing, part numbers, engineering change level, serial numbers, templates, and any special instructions were checked before assembly. The foil, which was a blueprint of component insertions, was placed on the light box. The foil contained the outline of two boards. The components were manually inserted one at a time, according to the foil. Upon completion of both cards, they were each verified with the CPL template. If any rework or scrap was to be done, it was performed in a manner similar to that of insertion. When the printed circuit boards were completely assembled and verified, they were manually placed into trays.

7.7.1 Measurement Method for Productivity

The task-oriented total productivity measurement (TOTP) model developed by Edosomwan (1985) was used in the case study. A brief description of TOTPM model is given below ($i = 1, 2, \ldots, n; j = 1, 2, \ldots, m;$ and $z = 1, 2, \ldots, k$):

$$
\begin{array}{l}
\text{Total productivity of task} \\
\text{i, in site j,} \\
\text{in period t}
\end{array}
=
\dfrac{
\begin{array}{c}
\text{Total measurable output} \\
\text{of task i, performed in site j,} \\
\text{in period t}
\end{array}
}{
\begin{array}{c}
\text{total measurable input of} \\
\text{task i, performed in site j,} \\
\text{in period t}
\end{array}
}
\qquad (7.1)
$$

where:

Total measurable output = value of (finished units produced + value of partial units produced + other output associated with units produced)

Total measurable input = value of (labor + materials + capital + energy + robotics expenses + computer operating expenses + other administrative expenses + data processing expenses)

Total-factor productivity is the ratio of total measurable output to the associated labor and capital input.

$$
\begin{array}{l}
\text{Total-factor productivity} \\
\text{of task i, in site j,} \\
\text{in period t}
\end{array}
=
\frac{\begin{array}{c}\text{total measurable labor and} \\ \text{capital output of task i,} \\ \text{performed in site j, in period t}\end{array}}{\begin{array}{c}\text{total measurable labor and} \\ \text{capital input of task i,} \\ \text{performed in site j, in period t}\end{array}}
\qquad (7.2)
$$

Partial productivity is the ratio of total measurable output to one class of input. For example,

$$
\begin{array}{l}
\text{Partial productivity of} \\
\text{task i, with respect to labor} \\
\text{input in site j, in period t}
\end{array}
=
\frac{\begin{array}{c}\text{total measurable output of} \\ \text{task i, performed in site j,} \\ \text{in period t}\end{array}}{\begin{array}{c}\text{measurable labor input of} \\ \text{task i, performed in site j,} \\ \text{in period t}\end{array}}
\qquad (7.3)
$$

The values of the output and input are expressed in constant monetary terms of a reference period. In the case study, the complexity factor of printed circuit boards and the proportional contribution to the total number of printed circuit board insertions was used as an allocation criterion for overhead expenses. Deflators were not required because the study period of 10 weeks was not long enough to be affected by significant price changes for input components.

7.7.2   Results

As shown in Figures 7.3 and 7.4, the implementation of the QER technique in a manufacturing task improved labor and total productivities. Perhaps the great success can be attributed to the teamwork approach adopted by the QC team, willingness to train each other for the skills required, and a strong ability to quickly reach a consensus to resolve problems at the source of production. Equally important was the high motivational level experienced by all team members. The feeling that the members of the team had total control over what they do and were able to take total responsibility for decision-making also contributed to the gains in productivity.

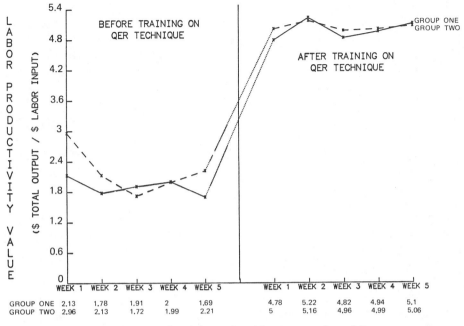

Figure 7.3   Labor productivity values for groups 1 and 2.

7.8   CASE STUDY: IMPLEMENTATION OF QER TECHNIQUE IN A BANK CLERICAL TASK

The quality error removal technique was implemented at a bank with specific attention to a check-processing task. This involved using a batch line processing approach to aid clerical workers in processing checks based on the priority of first in, first out. The task was performed manually. Before learning the QER technique, 37 members of the clerical staff performed this task. The labor productivity and defects per thousand checks processed were assessed without the training for the QER technique. The clerical staff members were then trained for the QER technique for a period of 8 weeks (on-the-job training accounted for 7 weeks and classroom education for 1 week).

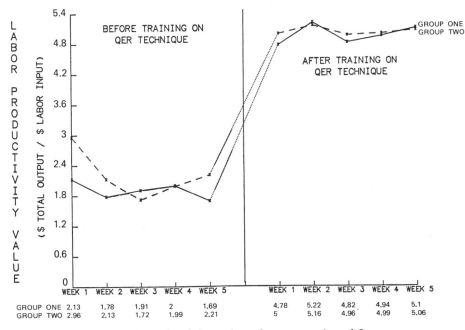

Figure 7.4   Total productivity values for groups 1 and 2.

7.8.1   Results

As shown in Figures 7.5 and 7.6, implementation of the QER technique in a bank check-processing task improved labor productivity and improved the quality of the work in the processing unit. The QER members were able to identify and resolve such problems as extra wasted transportation and motions that were nonvalue-add to their work. They were able to resolve bottlenecks that had an impact on the overall smooth flow between clerical personnel. The difficulty in processing each type of check was classified, and a solution strategy was developed and implemented.

7.9   PROBLEMS ENCOUNTERED DURING IMPLEMENTATION AND SUMMARY

The problems associated with the implementation of the QER technique may vary from one organization to another, depending on the na-

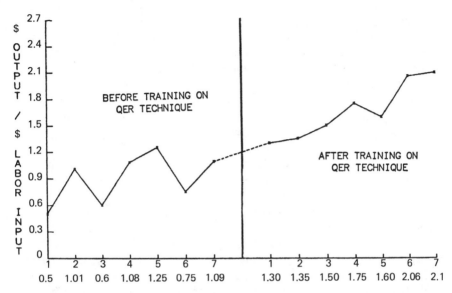

Figure 7.5   Labor productivity values for check processing task.

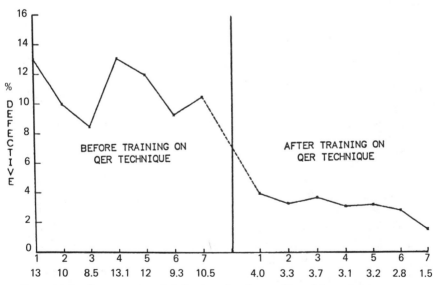

Figure 7.6   Percentage of defective checks attributable to human error in performing the check processing task.

ture of task performed, the QC team composition, and the levels of management support. Based on implementation in a manufacturing and a service task, it was observed that friction among team members arose in some situations. It is interesting that the team resolved all friction themselves.

It was also difficult for team members to accept and tolerate the rejection of ideas by management. This problem was resolved through ongoing communication between the QC team and management.

Problems also arose when the team had too many problems to resolve or too little to keep them busy. The team agenda improvement process helped resolve these problems.

As demonstrated by the case studies, the QER technique is a powerful tool for the improvement of productivity and quality at the source of production or service. The technique has a great potential for application in both the manufacturing and the service work environment. It is very inexpensive to set up and maintain the QER technique. QER basic strategy rests on providing decision latitude, control, and ownership to those charged with task responsibility. Management support, training, and other resources required are necessary for the successful implementation of the QER technique. The QER is unique and works best when the QC team is limited to a small group.

REFERENCES

Dewar, D. L., 1980. *The Quality Circle Handbook*. Quality Circle Institute, Red Bluff, California.

Doyle, M., and D. Straus 1978. *How to Make Meetings Work*. Playboy Paperbacks, New York.

Edosomwan, J. A., 1983. Quality Error Removal Technique. Unpublished Manual, IBM Data Systems Division, New York.

Edosomwan, J. A., 1985a. A task oriented total productivity measurement model for electronic printed circuit board assembly. Proceedings First International Conference on Electronic Assembly, October 7–9, 1985, Santa Clara, California.

Edosomwan, J. A., 1985b. A methodology for assessing the impact of computer technology on productivity, production quality, job satisfaction and psychological stress in a specific assembly task. Doctoral Dissertation, Department of Engineering Administration, The George Washington University, Washington D.C. 20052. Research sponsored by IBM Corporation and the Social Science Research Council (U.S. Department of Labor).

IBM Data Systems Division, 1982. *Teaming up for Quality: Quality Excellence Teams Education Guide.* Poughkeepsie, New York, 12602.

Robinson, M., 1982. *Quality Circle, A Practical Guide.* Gower Publishing, Aldershot, England.

8

Productivity and Quality Improvement Strategies and Techniques

A major challenge facing organizations is how to improve productivity and quality without having to make a major capital investment or increase the number of people required to perform a given task. In recent years, several "back to basics" techniques, such as the just-in-time (JIT), continuous flow manufacturing (CFM), stockless production, and process analysis technique (PAT), have received significant attention in the literature and in some cases in actual implementation. The concept of JIT has received more attention in Japan than anywhere else in the world. Schonberger (1982) details nine hidden lessons to be learned from the Japanese approach. Significant attention has also been focused on the application of robotic devices, computer-integrated manufacturing (CIM), computer-aided manufacturing, group technology, and other forms of techniques, such as process automation that eliminates defects. This chapter presents the factors affecting employee job performance and productivity and a discussion of key productivity and quality improvement techniques. Emphasis is placed on the production and service improvement technique (PASIT).

8.1 FACTORS AFFECTING EMPLOYEE JOB PERFORMANCE AND PRODUCTIVITY

Sutermeister (1973) identified various factors affecting employee job performance and productivity. These factors are shown in Figure 8.1.

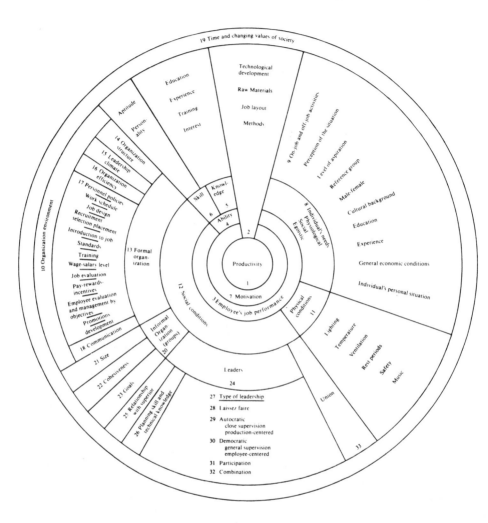

Figure 8.1   Major factors affecting employee job performance and productivity. Employees includes executives and managers, supervisors, professional staff, such as engineers and scientists, other white-collar workers, and blue-collar workers. (*Source*: R. A. Sutermeister, 1976. *People and Productivity*, 3rd Ed. McGraw-Hill, New York.)

The factors shown in each segment affect factors in the corresponding segment of the next smaller circle; they may also affect and be affected by other segments in the same circle or circles. Improving productivity will improve these various factors to favor continued productivity improvement as desired even by the employees themselves. However, the rise is not to be looked upon as self-generating or resulting from momentum. Productivity rises from the continuous application of human thought and effort to the end, not that employees work harder, but that they work better with better tools, materials, or methods. It is important to note that people in an organization give their best to the job to be done, not in a sustained and natural way, when the consequences of so doing are seen to be rewarding and meaningful. Tangible rewards—the shared gain of higher productivity—are part of a total pattern. Intangible consequences, too, yield satisfactions to which few people are indifferent: the pride of accomplishment, the zest of working with others in satisfying, productive effort, and the recognition of common ground and common goals. Improving the conditions of work, methods, tools, equipment, and good management will contribute to the productivity of employees.

8.2 PRODUCTIVITY AND QUALITY IMPROVEMENT TECHNIQUES

Sumanth (1984, pp. 318–319) reported 50 different techniques of productivity improvement that were obtained from a survey of the literature conducted by Omachonu (1980). These techniques are listed below:*

   I. Technology-based techniques
     A. Computer-aided design (CAD)
     B. Computer-aided manufacturing (CAM)
     C. Integrated CAM
     D. Robotics
     E. Laser beam technology
     F. Energy technology

Source: Sumanth, D. J., 1984. *Productivity Engineering and Management*. McGraw-Hill Book Company, New York, pp. 318–319). Reprinted with permission.)

    G. Group technology
    H. Computer graphics
    I. Emulation
    J. Maintenance management
    K. Rebuilding old machinery
    L. Energy conservation
II. Employee-based techniques
    A. Financial incentives (individual)
    B. Financial incentives (group)
    C. Fringe benefits
    D. Employee promotions
    E. Job enrichment
    F. Job enlargement
    G. Job rotation
    H. Worker participation
    I. Skill enhancement
    J. Management by objectives (MBO)
    K. Learning curve
    L. Communication
    M. Working condition improvement
    N. Training
    O. Education
    P. Role perception
    Q. Supervision quality
    R. Recognition
    S. Punishment
    T. Quality circles
    U. Zero defects
III. Product-based techniques
    A. Value engineering
    B. Product diversification
    C. Product simplification
    D. Research and development
    E. Product standardization
    F. Product reliability improvement
    G. Advertising and promotion
IV. Task-based techniques
    A. Methods engineering
    B. Work measurement
    C. Job design

    D. Job evaluation
    E. Job safety design
    F. Human factors engineering (ergonomics)
    G. Production scheduling
    H. Computer-aided data processing
V. Material-based techniques
    A. Inventory control
    B. Materials requirement planning (MRP)
    C. Materials management
    D. Quality control
    E. Material handling systems improvement
    F. Material reuse and recycling

8.3 PRODUCTION AND SERVICE IMPROVEMENT TECHNIQUE (PASIT)

Edosomwan (1983) developed the PASIT technique for improving productivity and quality in both service and manufacturing organizations.

8.3.1 PASIT Defined

According to Edosomwan (1983), PASIT is an ongoing process that involves the organized use of common sense to find easier and better ways of performing work and streamlining the production and service processes to ensure that goods and services are offered at a minimum overall cost. Improved productivity and quality is obtained through the elimination of waste, such as scrap, unnecessary automation, rework, inspection, excess work-in-process inventory, wasted motions and transportation, engineering changes to product specification, and other forms of waste that have no value added to the goods and services offered. The PASIT concept is shown schematically in Figure 8.2.

8.3.2 Benefits of PASIT

The implementation of PASIT in the work environment can provide the following benefits:

1. Optimal balanced service and production time is obtained with no waste.
2. Improve product and service quality and minimize inspection and repair loops.

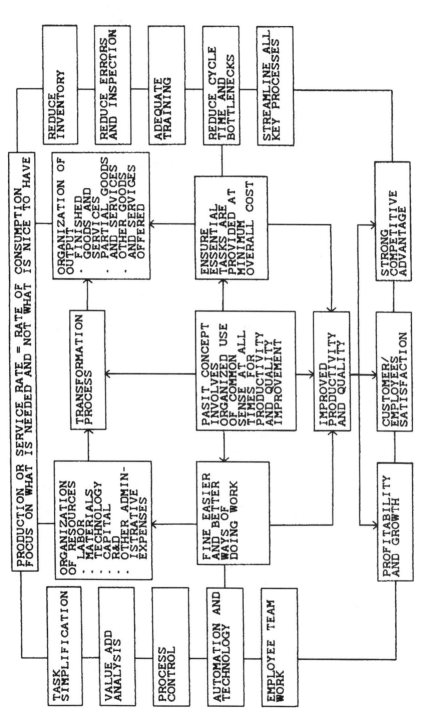

Figure 8.2   The PASIT concept.

3. Improve customer service time through reduction of improved service and product cycle time.
4. Reduce inventory through elimination of work-in-process at each service or production station.
5. Reduce process time for each task through elimination of unnecessary errors, steps, and handling defects.
6. Improve employee morale, job safety, and job satisfaction through elimination of repetitive work, service and production bottlenecks, reduction in training time, and reduction in the confusion and interdependencies among unnecessary functional tasks.

8.3.3 Major Principles Behind PASIT

In order for PASIT to provide the anticipated and stated benefits, the following seven principles must be adopted within the organization.

Principle 1. Management and employees must have the positive attitude that productivity and quality improvement can result from the organized use of common sense to address service and production problems. Management support must be shown through practice and an organization policy statement.

Principle 2. A total teamwork approach among functional organizations, such as research and development, marketing, personnel, purchasing, manufacturing, information systems, quality, facilities and distribution, maintenance, finance, production control, service centers, engineering, and other functions within the work organization, must be used to address all problems.

Principle 3. Total productivity and quality improvement at the source of production or service. Discourage heavy reliance on inspection and other nonvalue-add operations within the work organization. The required basic training must be provided to obtain good quality goods and services at the source of production or service.

Principle 4. Encourage reduction in the layers of management at all levels. Give ownership to those charged with task responsibility and the control and support needed to resolve daily service and production problems. Too many levels of management causes additional bottlenecks. A level of management must be instituted within the organization only if it will provide value-adding to the improvement of the production and service function.

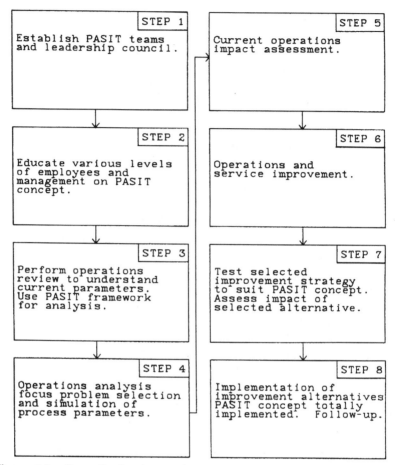

Figure 8.3   Steps for implementing the PASIT concept.

Principle 5. Total impact assessment for all service and process changes, policy changes, and implementation of new ideas and techniques. The new process, service, ideas, organization, or management must be better than the one it replaces, considering all implications.

Principle 6. Total reward system based on contributions in improving and managing all aspects of a task to obtain acceptable goods and services. Pay for performance in total technical and people management. Avoid a crisis loop reward system that affects the morale of the employees.

Principle 7. Production and service errors that affect productivity and quality are controllable through common sense and good judgment. The production rate must be equal to the consumption rate.

8.4 PASIT IMPLEMENTATION METHODOLOGY

The eight-step approach for implementing PASIT in a production and service organization is shown schematically in Figure 8.3. Each step will now be described.

8.4.1 Establishing PASIT Teams and Leadership Council

The PASIT teams should be made up of representatives from the working levels. Each functional area must be represented. The team members must possess, through training, the ability to resolve problems in a timely manner. They must be equipped with basic training about cause-and-effect relationships, priority mapping, value analysis, task simplification, process flow mapping, and other commonsense approaches to improving service or production environments. The PASIT team members should have the technical ownership for performing the variety of activities specified in Figure 8.4. The PASIT council manage-

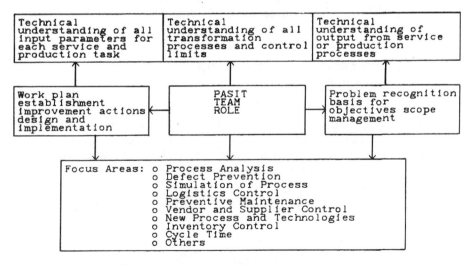

Figure 8.4   PASIT team members' roles.

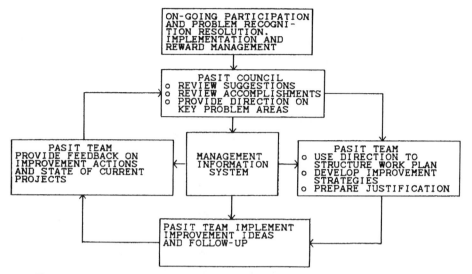

Figure 8.5   PASIT leadership council role.

ment role is to provide technical leadership, facilitate people involvement, review accomplishments, and clear communication up and down within the organization. The council should be made up of senior managers and specialists with training in managing productivity and quality improvement projects. The role of the council is shown schematically in Figure 8.5.

8.4.2   Educating Everyone on the PASIT Concept and Principles

Everyone involved with the task of implementing PASIT must be trained about the key principles and concepts. Four types of recommended session contents are presented in Table 8.1. The education sessions enable everyone to be thinking and talking about productivity and quality improvement in the same terms.

8.4.3   Understanding Current Service and Production Parameters

As shown in Figure 8.6, understanding current service and production parameters consists of six basic steps:

Table 8.1 Recommended PASIT Education Format

Level	Session duration (hr)	Recommended PASIT training content
Upper or senior management	4	Overview of PASIT technique and benefits; potential areas of application; PASIT principles
Middle management	8	PASIT concept and benefit; implementation requirements; how the concept works; managing PASIT projects; administering the reward from PASIT
First-line management, staff, specialist, engineers, and process analyst	24	Detailed concept and methodology for implementing PASIT; problem-solving framework; process-analyzing framework; simulation of processes; process improvement tools; improvement techniques and steps; justification and impact assessment tools; quality error removal technique; productivity improvement, measurement, planning, and evaluation tools
Operators, production specialists, service technicians	16	Implementing PASIT at work station; examples of PASIT applications; PASIT problem-solving framework; PASIT benefits in task improvement

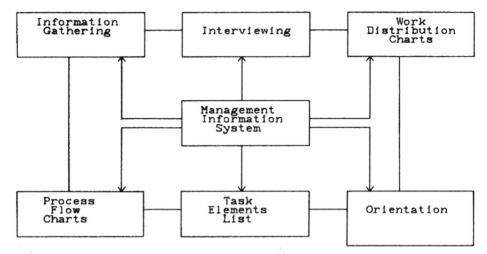

Figure 8.6   Elements involved in operations review.

1. Orientation to people, processes, and procedures
2. Information gathering
3. Interviewing
4. Preparing work distribution charts
5. Preparing process flow charts
6. Developing task element lists

The main accomplishment in this step should focus on the detailed review of tasks performed by individuals, work groups, departments, and functions; the methods used; and the flow of work from one task to another and among the members of the work group. It is also important to identify the volume of work and the frequency, rates, and timing involved in each task performed. In order to ensure adequate coverage of all key information, it is important to have a data-gathering checklist similar to that shown in Table 8.2.

Orientation

The orientation phase enables the analyst to know all the key processes, procedures, and people in the various organizations. The functions for each work group must be clearly understood. Attention must also be paid to interfunctional dependencies.

Table 8.2 Data Collection Checklist for Operations Review

Process item	Responsibility	Impact (usefulness)	Value adding or non-value adding
Task elements by key areas			
Procedures and policies			
Data-reporting format			
Organization charts			
Responsibilities by key areas			
Layout of work area			
Equipment or tools in use			
Cost tracking system			
Existing standards for work			
Forms in use			
Key processes			
Filing system			
Work load volume by key areas			
Yield indicators by key areas			
Production indicators by key areas			
Work assignment system			
Product or service distribution system			
Log books used in processes			
Work distribution charts			
Flowcharts for key processes			
Personnel names by key areas			
Organization rules and regulations			
Computer information type			
Cycle time of product or service			
Task distribution frequency			
Others			

Information Gathering and Interviewing

The objective is to get all facts on the task performed by asking ten basic questions:

1. What is it?
2. What does it do?
3. Who does it?
4. What does it cost?
5. Where it is being done?
6. What is it worth?
7. When is it done?
8. Why it is done?
9. Who has the authority to say so?
10. Who has ownership for it?

The interviewing process should be structured to obtain an opinion from every level of employees and management. The important point is to ensure that the analyst work with *facts*, not opinions. A scheduled time for interviews by a random selection process should also be encouraged.

Work Distribution Charts, Process Charts, and Task Elements

Work distribution charts, process flowcharts, and task element recording formats are organized ways of recording the information gathered from questionnaires, interviews, and process procedures. These charts enable analysts to identify the various task activities to be measured and to estimate the time spent by each individual or work group on each task. Flowcharts show the sequence of steps required to perform a task. Flowcharts also provide the basis for understanding the relationship between the task and the processing times. Task elements enable analysts to determine the extent of skill utilization, work specialization, delays, transportation required, and other details necessary to pinpoint and understand the process parameters.

8.4.4   Operations Analysis, Focus Problem Selection, and Process Simulation

In performing operations analysis, an analyst starts with a pilot task or product or service and obtains the following additional information:

1. Develop task flow diagram.
2. Record the processing times for each task.
3. Record setup time for each task.
4. Identify all transportation logistical data.
5. Simulate the parameters to understand input and output from each sector and leverage area. Simulation also identifies the impact of changes in processing times on other parameters.

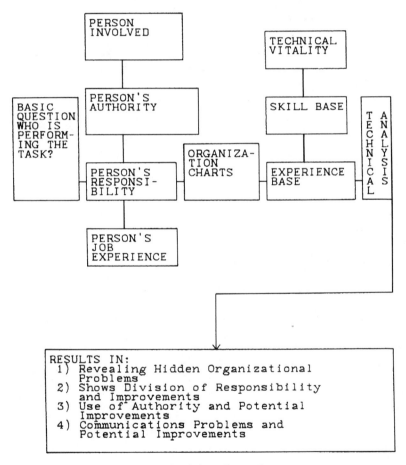

Figure 8.7   Analysis of *who* is doing the task.

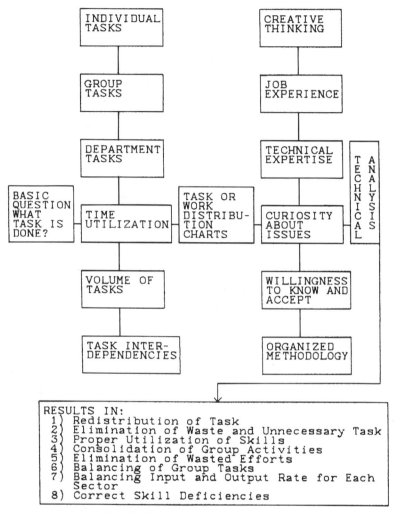

Figure 8.8   Analysis of *what* task is done.

The analysis phase provides the basis to determine bottlenecks, rework loops, capacity limitations, and resources constraints. The four major analysis results to be obtained by asking four key questions are presented in Figures 8.7 through 8.10.

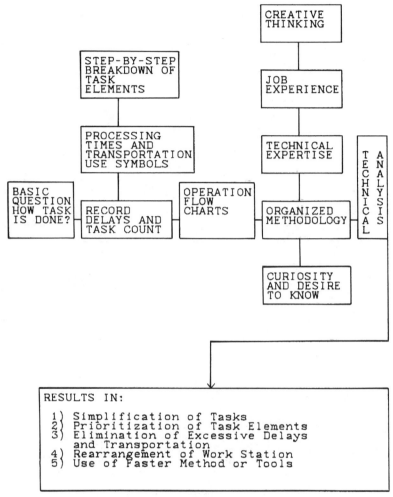

Figure 8.9   Analysis of *how* the task is done.

8.4.5   Current Operations Impact Assessment

In this step, the suitability of the current mode of operation is assessed. The feasibility and suitability of all parameters are determined based on cost, value-add, time, and impact on quality. Intangible elements,

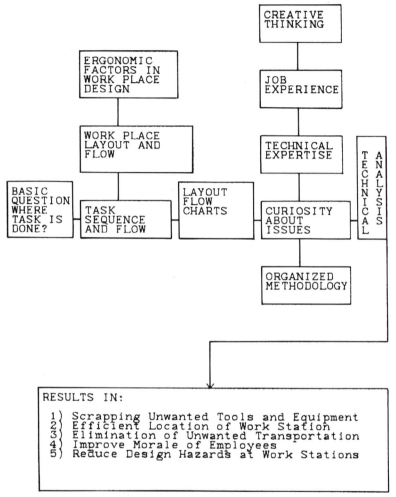

Figure 8.10   Analysis of *where* the task is done.

such as job satisfaction and morale, should also be taken into consideration.

8.4.6   Operations and Service Improvement

After completing the impact assessment of the current method, all new techniques and ideas for improvement should be formally documented. Each task must be carefully analyzed. In order to decide about a new improved method for operation, six general approaches should be considered:

1. Eliminate and minimize the number of task elements within a given operation.
2. Maximize the use of all resources available.
3. Combine and rearrange the sequences of processes.
4. Substitute and simplify methods of performing a given task.
5. Change the sequence for performing a given task.
6. Use a new technology or tool to replace the method of performing a given task.

8.4.7 Testing Selected Improvement Strategy

Different improvement strategies can basically be evaluated using the following criteria:

1. Cost and savings related to the specific alternative.
2. Quality improvement related to the specific alternative.
3. Job satisfaction and morale improvement related to the specific alternative.
4. Capability of the user to adapt to the specific alternative.
5. Implementation requirements of the specific alternative.
6. Time required to meet all objectives specified in the specific alternative.
7. Conformation of the specific alternative to established policy and standards.
8. Specific known exposures related to a specific alternative.
9. Justification of new alternative tools using the total productivity measurement approach.
10. Justification of a new alternative using cost-benefit analysis and other financial measures.

In situations in which intangible benefits outweigh tangible benefits, managerial judgment should be used to select the best alternative.

8.4.8 Implementation of Improvement Methods

The one key ingredient of the PASIT concept is that it rests on teamwork in all phases. In order to implement new ideas, the approval of all key functional representatives and management must be obtained. The new method must be reviewed at all levels and revised based on useful suggestions. Specific tasks should be assigned to specific people to implement the aspects of the new alternative. Implementation of the PASIT concept and approach is not a one-time action; it is an ongoing process for production and service improvement as long as the organization exists.

8.5 BASIC TOOLS NEEDED FOR PASIT IMPLEMENTATION

8.5.1 Processing Time

Processing time is the amount of time required to perform a given task. The usual source for this time is the time standards prepared by industrial engineers.

8.5.2 TAKT Time

TAKT time is the amount of time required to perform each task and obtain the total volume required by a sector within a specified time period.

$$\text{TAKT time} = \frac{\text{minutes available per day—lunch, breaks}}{\text{daily going rate (volume)}}$$

8.5.3 Capacity

This is a measure of how much a given production or service unit can offer given all the resources available at the unit in a specific time period.

8.5.4 Utilization

This is a measure of how well people, tools, equipment, materials, and other resources are utilized.

8.5.5 Yields

This is a measure of the quality of the output expressed in defects per unit or the percentage of defective material in a sample.

REFERENCES

Edosomwan, J. A., 1983. Production and Service Improvement Technique (PASIT). Unpublished manual. IBM Data Systems Division, New York.

Edosomwan, J. A., 1985. A methodology for assessing the impact of computer technology on productivity, production quality, job satisfaction, and psychological stress in an assembly task. Doctoral Dissertation, Department of Engineering Administration, The George Washington University, Washington, D.C. 20052. Research sponsored by Social Science Research Council (U.S. Department of Labor) and the IBM Corporation.

Edosomwan, J. A., 1986a. Productivity management in computer-aided manufacturing environment. Proceedings of First International Conference on Engineering Management, September 22–24, Washington, D.C.

Edosomwan, J. A., 1986b. Technology impact on the quality of work life—a challenge for engineering managers in the year 2000. Proceedings of First International Conference on Engineering Management, September 22–24, Washington, D.C.

Kendrick, J. W., in collaboration with the American Productivity Center, 1984. The John Hopkins University Press, Baltimore, Maryland.

Mali, P., 1978. *Improving Total Productivity: MBO Strategies for Business, Government, and Not-For-Profit Organizations*. John Wiley, New York.

Omachonu, V. K., 1980. Productivity improvement: Conceptual framework, model and implementation methodology for manufacturing companies. M.S. Thesis, Department of Industrial Engineering, University of Miami, Coral Gables, Florida, July.

Schonberger, R. J., 1982. *Japanese Manufacturing Techniques: Nine Hidden Lessons in Simplicity*. Free Press, New York.

Sumanth, D. J., 1984. *Productivity Engineering and Management*. McGraw-Hill, New York.

Sutermeister, R. A., 1976. *People and Productivity*, 3rd Ed. McGraw-Hill, New York.

9

Job Simplification and Motivation, and Morale Improvement

This chapter presents other techniques that will help in productivity and quality improvement. An overview of work measurement and a job simplification program are provided. Techniques for motivation and morale improvement are discussed. The impact of technology and task design on productivity and quality is also presented.

9.1 WORK MEASUREMENT AND JOB SIMPLIFICATION PROGRAM

One of the objectives to be met when attempting to increase individual productivity is a better set of measurements of activities. In order to deal with improvements there must be bench marks and yardsticks in the form of indexes, standards, and units for all activities.

Defined in its broader sense, work measurement is a means of determining an equitable relationship between the quantity of the work performed and the number of labor hours required for completing that quantity of work. The establishment of measurement is for the purpose of planning, scheduling, and controlling work. Organizations should tailor their staff functions to have work measurement techniques that will help to achieve the following objectives.

Acquaint staff personnel in charge with more formal methods of productivity management and labor performance.

The standard developed by work measurement should include standard allowances for rest, delays that occur as part of the job, time for personal needs, and, where the work is heavy, an allowance for personal fatigue.

Employees should be involved. They should be shown the part to play if formal methods are to be used successfully.

All participants should be given the confidence that the formal methods do work and should be used.

If carefully implemented, a work measurement program will help an organization to achieve the following:

A tool for decision-making

Data provided for scheduling day-to-day operations

A basis for labor cost control

The method of doing work must be subjected to review and study, with a view of making the process simpler, more nearly perfect, and more productive—these can be termed work simplification.

Work simplification is the systematic investigation and analysis of contemplated and present work systems and methods for the purpose of developing easier, quicker, less fatiguing, and more economic ways of providing high-quality goods and services. Consideration is given to improving the product or service, raw materials and supplies, the sequence of operations, tools, work place, and equipment and hand and body motions. It is a systematic approach of defining, analyzing alternatives, and documenting work methods. Some of the techniques used for work measurement and job simplification are:

1. Process charts and flow diagrams
2. Operation charts
3. Micromotion study
4. Principles of motion economy
5. Job simplification
6. Time standards
7. Value analysis
8. Operation research techniques
9. Human factors
10. Standard data (MTM)
11. Rating schemes

12. Stopwatch method
13. Work sampling

A more detailed theoretical discussion of these techniques can be found in the works of Barnes (1968) and Mundel (1970). The implementation of a work measurement and job simplification program should include a brainstorming session that provides an opportunity for input from all employees.

9.2 MANAGEMENT DEVELOPMENT PROGRAM

There has been a growing trend in organizations to focus on external management development activities. This can be attributed to the fact that there is growing dissatisfaction with the results obtained from off-the-job management training programs. The dissatisfaction arises because these programs are not tailor-made to the individual organization and they do not bring about the expected changes in managerial behavior. Also, the manager who would improve individual productivity must select with extreme care the people who are to supervise the workers. Those selected must be carefully and continuously trained in the modern concepts of supervision and management. The truculent, ignorant, untrainable, and psychologically disturbed should never be allowed to supervise the work of other people, yet many organizations select supervisors by the most random methods, train them not at all, and wonder why productivity does not evolve.

The educational function of the management can be used to establish a climate of receptive attitudes toward change and to facilitate communications between employees. Change is not always welcomed by even the most conscientious supervisor or worker. No matter now capable the manager, he or she will encounter resistance to change. Resistance to change, a well-established psychological trait, thrives on fear and apprehension and is nourished by poor communication. The effective manger is well trained and overcomes resistance to change through effective communication. Communication and the establishment of good rapport can be greatly helped by acquainting the manager with terminology and techniques.

There have been instances in which a lack of understanding can lead to total rejection of a new concept. Many managers will not accept what they do not understand and appreciate. A manager must be made aware of the approach to problem solving so that the manager can appreciate a new system design or improved method.

If the management development programs conducted are to achieve their purpose, the subject matter must acquaint participants with well-explained principles and techniques. Management development should the type of policies to institute, when to institute them, and how. Some examples of the areas for management training are:

1. Basic fundamentals of management directing, planning, staffing, controlling
2. Managing the leveled employee
3. Employee development guidelines
4. Company operating policies
5. Motivation techniques
6. Problem-solving techniques
7. Productivity and quality improvement techniques
8. Salary and benefit administration
9. Handling grievances, sick leave, income, disabled employees, job seniority, job training, job posting, layoffs, jury duty, and others
10. Investment strategies and competition

Mali (1978, pp. 173–177) discusses six key supervisory actions for developing productivity mindedness:

1. Supervisors must give facts and information to counter negative attitudes.
2. Supervisors themselves must be examples of productivity mindedness.
3. Give productivity orientation and training to employees.
4. Communicate concern and the need for productivity improvement to employees.
5. Allow participation in productivity decision-making.
6. Use organized programs.

Goodwin (1968, pp. 530–543) also pointed out the need for a people-oriented approach to improvement management. He emphasized that "people are our most important asset and their attitudes and motivational drives are a major factor in any successful business." The improvement management framework suggested by Goodwin is shown in Figure 9.1.

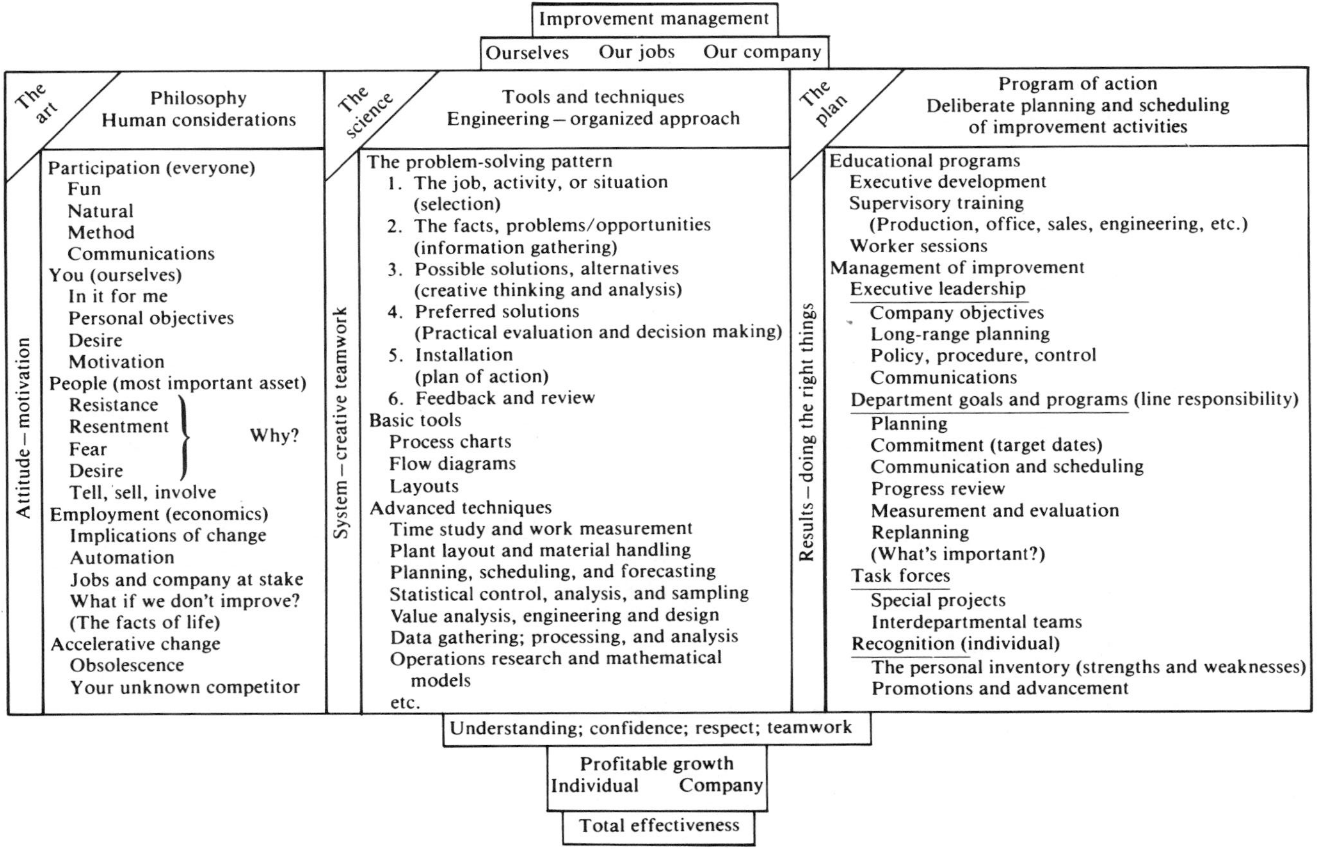

Improvement management

Ourselves   Our jobs   Our company

The plan

Program of action
Deliberate planning and scheduling
of improvement activities

Educational programs
  Executive development
  Supervisory training
    (Production, office, sales, engineering, etc.)
  Worker sessions
Management of improvement
  Executive leadership
    Company objectives
    Long-range planning
    Policy, procedure, control
    Communications
  Department goals and programs (line responsibility)
    Planning
    Commitment (target dates)
    Communication and scheduling
    Progress review
    Measurement and evaluation
    Replanning
      (What's important?)
  Task forces
    Special projects
    Interdepartmental teams
    Recognition (individual)
  The personal inventory (strengths and weaknesses)
    Promotions and advancement

Results — doing the right things

The science

Tools and techniques
Engineering — organized approach

The problem-solving pattern
  1. The job, activity, or situation
     (selection)
  2. The facts, problems/opportunities
     (information gathering)
  3. Possible solutions, alternatives
     (creative thinking and analysis)
  4. Preferred solutions
     (Practical evaluation and decision making)
  5. Installation
     (plan of action)
  6. Feedback and review
Basic tools
  Process charts
  Flow diagrams
  Layouts
Advanced techniques
  Time study and work measurement
  Plant layout and material handling
  Planning, scheduling, and forecasting
  Statistical control, analysis, and sampling
  Value analysis, engineering and design
  Data gathering; processing, and analysis
  Operations research and mathematical
    models
    etc.

System — creative teamwork

The art

Philosophy
Human considerations

Participation (everyone)
  Fun
  Natural
  Method
  Communications
You (ourselves)
  In it for me
  Personal objectives
  Desire
  Motivation
People (most important asset)
  Resistance   Why?
  Resentment
  Fear
  Desire
  Tell, sell, involve
Employment (economics)
  Implications of change
  Automation
  Jobs and company at stake
  What if we don't improve?
  (The facts of life)
Accelerative change
  Obsolescence
  Your unknown competitor

Attitude — motivation

Understanding; confidence; respect; teamwork

Profitable growth
Individual   Company

Total effectiveness

Figure 9.1 Model for "Improvement Management." (*Source:* Goodman, H. F., 1968. Improvement must be managed. *Industrial Engineering,* Vol. 19, No. 11, pp. 538–543. Reprinted with permission.)

9.3 MOTIVATION OF EMPLOYEES

For several decades, productivity and quality improvement have been linked to how well employees are motivated to do their job. At the organizational level, several attempts have been made to arrange job conditions to enhance the coincidence of needs and satisfaction for a worker and the organization. In a highly technologically complex environment, motivating workers is difficult because motives have become more complex and diffused. A person's job is more than an opportunity to earn money with which to satisfy physical wants and needs. The need now also includes, among others, social experience that contributes to satisfy other noneconomic wants and needs brought to the job. Different organizations with different tasks, different competitive environments, and different worker personal needs require different approaches to motivating.

In general, motives are viewed as reasons for behavior. They direct people toward goals and affect both their ability and their will to perform the job. Motivation can be defined as motive strength to satisfy a need. Since human behaviors are complex, what motivates individuals is therefore based primarily on certain assumptions about human behavior. The three forms of motivation widely used are presented in Table 9.1. Some of the motivation techniques proposed in the literature are summarized in Table 9.2. In this high-technology era, the way jobs and tasks are designed and the opportunities that arise from such design may be one of the most substantial influences on the motivation of workers and the productivity of individuals and the organization. Job design practices should be based on a general approach that will provide opportunities for individual need satisfaction and goal achievement.

9.4 MOTIVATING WORKERS IN A HIGH-TECHNOLOGY PRODUCTION ENVIRONMENT

During the latter part of the nineteenth century, a new scientific management had tremendous influence on the production environment. Taylor, Gilbreths, and Gantt pioneered an approach that required job specialization and piece rate as the means of payment of wages. The excessive job specialization and machine-paced jobs and

Table 9.1   Three Approaches to Motivation

Items	Traditional economic incentive approach	Human relations approach	Self-drive approach
Significant managerial assumptions	1. People work only for money 2. People have no self-direction 3. People need leadership 4. People dislike work 5. Economic incentive will keep people on the job to perform	1. People like to work 2. People need comfortable work environment 3. Need for good leadership 4. Happy employees will produce more 5. Happy employees will be committed to the organization	1. People want meaningful work 2. Individuals possesses self-direction to work 3. People have self-control 4. Creative mind
Significant rewards offered	1. Good salary 2. Economic awards	1. Managerial recognition 2. Good salary 3. Good working condition 4. Good fringe benefits 5. Peer recognition	1. Personal growth 2. Opportunity to use skill 3. Achievements 4. Recognition 5. Good salary and benefits
Methods for using rewards to motivate	1. Piecework system 2. Simple tasks 3. Close supervision while performing task 4. Clear instructions to perform task 5. Focus on salary administration	1. Good management 2. Good organization policies 3. Fringe benefits 4. Teamwork 5. Considerate supervision	1. Autonomy 2. General supervision 3. Open system career policies 4. Goal setting 5. Job content
Significant needs satisfied	1. Physical need 2. Satisfaction from wealth acquired	1. Social need 2. Security need	1. Egoistic need
Impact	1. Job dissatisfaction 2. Lack of total participation and commitment	1. Productivity increase from satisfaction 2. Participative management approach in most issues	1. Productivity results brings additional satisfaction 2. Participative management in major issues

Table 9.2 Summary of Some of the Motivation Techniques

Author	(Year)	Motivation technique
Smith	(1776)	Motivate workers through financial incentives; strong enough to demonstrate monetary gains
Taylor	(1911)	Motivate workers through division of labor; divide work into standards, assign pay to standards, and give an easy procedure to reach standards
Munsterberg	(1913)	Motivate workers through job structure steps, assign money rates to the steps, and provide merit evaluation when it is recognized these jobs have been achieved
Maslow	(1943)	Motivate workers by providing opportunities that fulfill their needs (physiological needs, safety needs, love needs, esteem needs, and self-actualization needs)
Herzberg	(1959)	Motivate workers by providing job content that will lead to satisfaction
McGregor	(1960)	Motivate workers by giving managers a new view and set of attitudes about people and their environment for best advancing goals of people and the organization: two sets of theories, X and Y
McClelland	(1961)	Motivate workers by providing entrepreneurial development opportunities to experience high levels of achievement
Vroom	(1964)	Motivate workers by providing, in advance, opportunities in which rewards are great and the probability of achieving them is high
Mali	(1972)	Motivate workers by planning and obtaining the closest possible alignment between employee expectancies and organizational objectives
Edosomwan	(1985)	Motivate workers by providing meaningful job content, autonomy, ownership for results, and opportunities to fulfill their needs and that of the organization; reward all accomplishments in a timely manner

the use of piece rates can be viewed as a temporary source of motivation. In a rapidly changing technological environment, the individual source of motivation is more than the opportunity to earn money or work on a specialized task. Motivational factors should focus on both individual and company goals. The job should be enlarged and enriched in the following two ways.

9.4.1 Total Vertical Job Enlargement and Enrichment

The total vertical job enlargement and enrichment (TVJEAE) process involves both employees and management in planning, organizing, performing, and improving the job content. Technology is viewed and used under this situation as a mechanism to aid the production worker, not as a mechanism to control the worker. The TVJEAE approach believes in providing workers with total ownership and control over their daily activities. Workers have latitude in making decisions to improve job aspects that are impediments to their overall effectiveness. Such items as machine speed can be adjusted by the worker to a comfortable pace that permits overall satisfaction and productivity and quality improvement. the TVJEAE brings most motivations into play. Workers that have problems with such issues as technology user friendliness, pace, and task specialization have the opportunity to work with management to resolve such issues under the TVJEAE concepts.

9.4.2 Total Horizontal Job Enlargement and Enrichment

The total horizontal job enlargement and enrichment (THJEAE) process involves both employees and management in expanding job responsibilities to include a greater variety of activities. THJEAE aims at counteracting oversimplification and gives the worker an opportunity to perform a more expanded work unit. THJEAE also helps to eliminate monotony, a repetitive boring job, and specialization that may be caused by technology applications.

In a technology production environment there are several factors that strongly determine job satisfaction. Some of the key factors are job content, employee decision latitude, responsibility, advancement, recognition, employment security, and achievement. Working conditions, salary, company policy and administration, and management programs are other equally important hygiene factors. A comprehensive motivation strategy in a technology production environment should rest on the premise that

Technology is a mechanism design to aid the worker, not to control the worker.

There is freedom to use imagination and ingenuity in performing a task.

The opportunity exists for growth through meaningful assignment.

The opportunity exists to learn new skills and broaden a background on an ongoing basis.

The opportunity exists to earn more money in recognition of achievement.

The opportunity exists to maintain full employment and participate in the decision-making process.

The elements involved in motivating workers in a high-technology productivity environment are shown schematically in Figure 9.2.

9.5   MORALE MANAGEMENT

The morale of employees and management can have an impact on the productivity and quality of the work performed. Individuals with good feelings and attitudes toward the job are bound to have a positive impact on improving the overall effectiveness of the set of activities addressed. The employee and management morale management program (EMMMP) should focus on ongoing communication between management and employees. Communication should be geared toward revealing issues of importance to both employees and management that affect their morale. A total team effort should then be used to provide actions that are tailored to address individual concerns. Some companies have a formal morale assessment program that provides the basis for areas that need action plans. If daily ongoing communication is done properly, a formalized morale assessment is really not necessary. A key ingredient in EMMMP success is the ownership of morale improvement by the management team. Each manager's emphasis should be an honest attempt to understand employee morale through informal and formal discussions, through group discussions, and through a participative management approach in performing daily

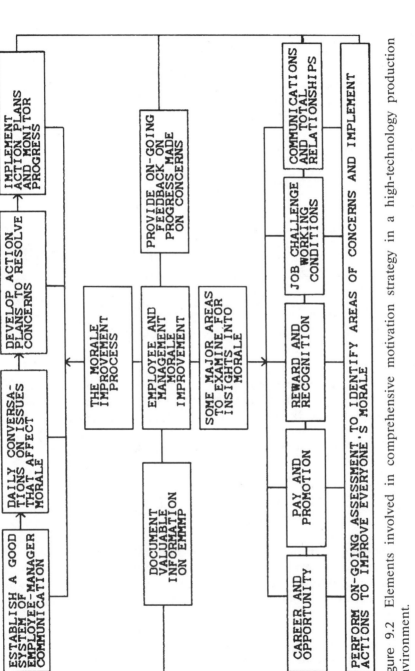

Figure 9.2 Elements involved in comprehensive motivation strategy in a high-technology production environment.

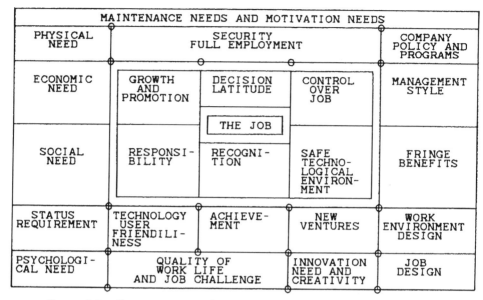

Figure 9.3   Components and process of EMMMP.

duties. It is recommended that ongoing feedback sessions be conducted between employees and management regarding matters of concerns. A strategy that works well is the continuous reinforcement of positive behaviors and accomplishments. Positive feedback on problems resolved also shows that someone cares. The EMMMP components and process are shown schematically in Figure 9.3.

9.6   TECHNOLOGY FACTOR IN PRODUCTIVITY AND QUALITY IMPROVEMENT

Technology can be defined as a mechanism for transformation in a production or service process. It has the capability of affecting all levels of contact in human endeavor. Between 1960 and 1986, technology has primarily been a major factor that has a significant impact on productivity and the overall quality of working life. Recent studies conducted by Edosomwan (1985) showed that in areas in which robotics and computer-aided manufacturing techniques are applied, productivity and quality improved significantly (see Table 9.3).

Table 9.3   Summary of Specific Type of Technology Impact on Productivity and Quality

Types of technology	% Improvement in total productivity	% Improvement in quality
Robotics devices	47	29
Computer-aided devices	53	21
Group technology	68	31

Morrison and McKee (1978) also reported some expert views on the contribution of technology to an increase productivity. As shown in Table 9.4, technology was reported to be the greatest contributor to the productivity increase.

Some of the many reasons that technology has a tremendous impact are as follows:

1. Provide or create ability to produce more in a shorter time cycle
2. Provide means to reduce quality errors
3. Provide a logical approach to assist human reasoning in completing tasks
4. Provide the ability for consistency in processes and services when applied
5. Provide ability for self-generating, uninterrupted work stations

Table 9.4   Comparison Among Labor, Capital, and Technology Impact on Productivity

	Factors		
Experts	Labor (%)	Capital (%)	Technology (%)
Kendrick	10	18	72
Denison	10	20	62
Christenson, Cummings, and Jorgenson	14	42	44
Average	14	27	59

Source: Morrison, D. L., and K. E. McKee, 1978. Technology for improved productivity. Manufacturing Productivity Frontiers, Vol. 2, No. 6, June, pp. 1–6.

Although such technologies as robotics, CAD/CAM, laser, and group technology are highly recommended as productivity and quality improvement tools, they must be justified properly using the methods discussed in Chapter 2. The total cost of doing business must be taken into account when assessing their impact on total productivity.

9.7 DESIGN IMPACT

Productivity and quality can be greatly influenced for better or worse by the physical well-being of tools, technology, and work places. Edosomwan (1986, p. 18) provided the following seven design rules that can help emerging technologies and new work places.

Rule 1. The individual is the center of design. Recognize the human anatomic structure, and obtain anthropometric dimensions to get the task to the individual.

Rule 2. Utilize the principles of kinesiology in the design. Avoid incompatible movement design, and allow the free movement of joints around joint axes.

Rule 3. Observe the individual's physiological capacity; avoid static postures and physiological responses as criteria for design, and avoid stress concentration points.

Rule 4. Apply psychological principles to improve morale and increase job satisfaction.

Rule 5. Recognize worker rights to the control of production of work activity, and allow skills development.

Rule 6. Involve the users of technology-based systems in the design phase.

Rule 7. Give technology-based systems the ability to advise, alert, or warn users of potential events.

9.8 OTHER FACTORS THAT IMPROVE MORALE

Although, there are no easy approaches to improving morale, the following actions can help in an on-going program to improve morale in the workplace:

1. On-going and frequent departmental meetings to enhance communication between employees and management.

2. On-going management of quality improvement sessions to resolve business problems.
3. Timely recognition of achievements and job performance.
4. On-going development of employee's talents, both managers and non-managers.
5. On-going management attention to problems in the workplace.
6. An on-going open door policy that enables employees and management to communicate frequently concerning workplace problems.
7. On-going interfunctional communication to resolve problems.
8. Emphasis on people management to resolve any issue that might affect morale.

REFERENCES

Barnes, R. M., 1968. *Motion and Time Study Design and Measurement of Work*, 6th Ed. John Wiley and Sons, New York.

Edosomwan, J. A., 1986a. Managing technology in the work place—a challenge for industrial engineers. *Industrial Engineering*, February.

Edosomwan, J. A., 1986b. A methodology for assessing the impact of robotics on total productivity in an assembly task. Proceedings of Annual International Industrial Engineering Conference, May, Dallas, Texas.

Goodwin, H. F., 1968. Improvement must be managed. *Journal of Industrial Engineering*, Vol. 19, No. 11, pp. 538–543.

Herzberg, F., B. Mansner, and D. B. Snyderman, 1959. *The Motivation to Work*. John Wiley and Sons, New York.

Mali, P., 1972. *Managing by Objectives*. John Wiley and Sons, New York.

Maslow, A. 1943. A theory of human motivation. *Psychological Review*, July, pp. 388–389.

McClelland, D. C., 1961. *The Achieving Society*. Van Nostrand, Princeton, New Jersey.

McGregor, D., 1960. *The Human Side of Enterprise*. McGraw-Hill, New York.

Morrison, D. L., and K. E. McKee, 1978. Technology for improved productivity. *Manufacturing Productivity Frontier*, Vol. 2, No. 6, June, pp. 1–6.

Mundel, E. M., 1970. *Motion and Time Study Principles and Practice*. Prentice Hall, Englewood Cliffs, New Jersey.

Smith, A., 1776. *The Wealth of Nations*.

Taylor, F. W., 1911. *Scientific Management*. Harper and Brothers Publishers, New York.

Vroom, V. H., 1964. *Work and Motivation*. John Wiley and Sons, New York.

Walker, R. C., 1950. The problem of the repetitive job. *Harvard Business Review*, Vol. 28, No. 3, pp. 54–58, May.

10

Establishing and Managing a Productivity and Quality Improvement Program

This chapter presents guidelines for setting up and managing productivity and quality improvement programs in organizations. A strategy for resolving productivity and quality problems is presented. A formalized approach for overcoming common concern is also offered. A checklist to be used by organizations in self-assessment is provided.

10.1 KEY ELEMENTS NECESSARY FOR A SUCCESSFUL PRODUCTIVITY AND QUALITY PROGRAM

Edosomwan (1986) discusses ten key elements required for successful productivity and quality programs. The elements are presented in Figure 10.1. Each step will now be described.

10.1.1 Top Management Support

A successful productivity and quality improvement program requires the strong commitment and involvement of top executives and senior managers of the organization. Total commitment from top management should be demonstrated through policies, practice, and support provided to have a formalized organizational structure that is in charge of productivity and quality improvement programs. The commitment

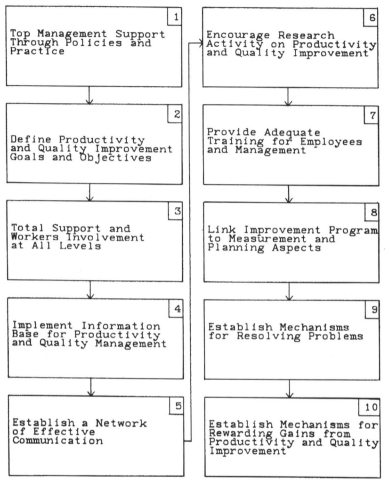

Figure 10.1   Elements involved in promoting a successful productivity and quality improvement program.

must state that no levels of defect are acceptable. Emphasis should be placed on productivity improvement through defect and error prevention programs. Inspecting quality into the product or service should be discouraged. The commitment should be to do it right the first time at the source of production or service. Top management involvement requires the ongoing participation review of productivity and quality improvement projects and productivity leadership to resolve controversial issues and allocate resources for productivity and quality improvement programs.

10.1.2 Define Productivity and Quality Improvement Goals and Objectives

At both the basic task level and the organizational level, there should be a clear definition of productivity and quality improvement goals. This provides the basis for dealing with complicated and deep-rooted problems. Pareto principles should be used to classify goals and problems when applicable. Each productivity and quality goal and problem should be matched with implemented action on a real-time basis. The projected benefits should be stated and monitored.

10.1.3 Worker Involvement

A participating management approach should be used to involve workers at all levels. Everyone should be made to understand the meaning of quality and productivity in the survival contest for profit. The commitment from both management and workers should be to do it right the first time at the source of production or service. Teamwork should be encouraged. A productivity and quality improvement team should include members from each sector of the production or service process. The team should meet regularly to resolve potential problems. Brainstorming techniques should be encouraged at the team level. Workers should also be involved in a goal-setting process; this will enable them to take ownership for the productivity and quality issues.

10.1.4 Implement Information Base for Productivity and Quality Management

A management information system should be implemented. Such a system should incorporate a data base on productivity and quality trends, issues of key accomplishments, and measurements and plans. The data base of key processes, services, and procedures should be made available to analysts and team members working on the improvement of productivity and quality.

10.1.5   Establish a Network of Effective Communication

At all levels of the organization, people should be made aware of productivity and quality issues. Such communication can be done through special reports on productivity and quality, bulletins, meetings set up to review issues, departmental meetings, and management memoranda. It is also important for the productivity coordinators and analysts to keep abreast of the latest developments in the area through external journals, conferences, and library research. Participative management approach should be used at all levels. This will improve communication up and down the organization.

10.1.6   Encourage Research Activity on Productivity and Quality Improvement

Productivity and quality improvement result from both the ability to generate new ideas and techniques and the ability to apply such new ideas to improve current operations. An organization that provides both financial and moral support to research activity is bound to be profitable. The encouragement of research activity for the improvement of productivity and quality also helps create a fertile climate for innovation of new products, new services, and new techniques of management.

10.1.7   Provide Management and Employee Training for Improving and Directing Productivity and Quality

Everyone in the organization should be made to understand the goals and objectives of the productivity and quality improvement program. This will facilitate commitment from everyone. At all levels, specific training should be designed to educate people on productivity and quality improvement concepts. A typical training package should include the following: techniques for productivity and quality measurement, control and evaluation, planning and analysis, improvement and monitoring, root-cause analysis technique, design of experiments, project and task management, and construction and interpretation of controls. Training provided should require tools and awareness of the issue of quality and productivity.

10.1.8 Link Improvement Program to Measurement and Planning Aspects

A total productivity measurement system should be implemented at both the task and the firm level. Such a system enables an organization to relate partial productivities to total and to pinpoint specific areas for improvement. Institute a statistical process control technique when possible. Develop measurements for percentage defectives, rework, and engineering changes, and measure the variability of the process on an ongoing basis. The process control charts should be understandable by everyone as a measure of process effectiveness. The cost of quality and productivity improvements implemented should be calculated periodically. Actions should be implemented to correct high failure rates and errors at the source of production or service. Short-term and long-term planning for zero defects and productivity gains should be developed.

10.1.9 Establish Mechanisms for Resolving Problems

In addition to encouraging teamwork at all levels, organization leadership may elect to use experienced managers and technical personnel, consultants, and a task force to resolve productivity and quality issues. The use of individual efforts, such as experienced technical personnel within the organization, allows those most intimately familiar with the current work processes to suggest and implement improvements. The danger in using this approach is that people may not want improvement in the processes within their realm of authority. It also may be difficult to see improvement within a given process, given that participants have been used to performing the same functions the same way for a long period of time. Organizations that choose to use this approach must be willing to share gains with their employees and managers on new ideas about productivity and quality management.

The second alternative is to use the task force when applicable. The task force approach has advantages, such as reduced training cost, uses most competent people, increases objectivity from members, and is often comprised of members from all functions and sectors. The drawback to this approach is that a commitment for follow-up on improvement projects could be missing after the task force is dissolved. The way to avoid this is to assign a specific department or person for full accountability of improvement projects provided by the task force.

The third alternative is to use a consultant. This approach brings strong technical proficiency if the consultant is selected properly. The consultant often has neutral opinions about organization policies and has one key mission of doing the best job for fair pay. This approach often provides honest, fresh ideas and does not require the internal training of workers. However, it might be pointed out that the end product obtained from a consultant is only as good as the qualification and experience of the consultant. Consultants are often resisted by insiders. A strong effort must be made to seek his or her ideas for improvement. A major drawback to using a consultant is that the burden of commitment and accountability may remain hanging with the senior management of the organization unless it is delegated quickly. The vast majority of consulting firms are honest and reliable. To protect organizational assets against the fraudulent few, the following should be done:

1. Request background information on the consulting firm and leading experts.
2. Find out if there is charge for estimates before you request one.
3. Find out if there is a charge for proposals or a site visit before you request one.
4. Beware of high-pressure improvement promises at low cost.
5. Deal with consulting firms that are well known or recommended by peers or experts. Consult productivity and quality improvement centers when in doubt (see Appendix E).
6. Allow some internal employees to work with consultants to facilitate easy implementation of new ideas and consultant understanding of processes and procedures.
7. Beware of unusually low prices for services that seem too good to be true; they probably are.
8. Ask for a detailed project schedule and implementation plan for action.
9. Understand how estimates for resources can change and the associated cost.
10. Ensure that the work done, improvement actions suggested, organizational charges, and other recommendations and followup actions are provided.

10.1.10 Establish Mechanisms for Rewarding Gains from Productivity and Quality Improvement

Ideas implemented for productivity and quality improvement should be rewarded by management through several means, such as awards, recognition among peers, bonus pay, additional technical challenge, and promotion. Management should ensure that compensation and promotion should be related to good performance.

10.2 ORGANIZATION FOR PRODUCTIVITY AND QUALITY MANAGEMENT

In Chapter 1, the productivity and quality management hierarchy (PQMH) was presented. At each level of the hierarchy, it is important to specify the organization structure with the individuals responsible for the improvement program. At the organizational level, the productivity and quality improvement director, coordinators, and program managers are usually selected to hold such activity. At the policy level, a productivity steering committee is usually appointed to provide leadership for the organization's productivity and quality goals and objectives. At lower levels, such as by department or plant, a productivity council should also be formed to promote improvement activities. A summary of the role and responsibilities of the steering committee, coordinator, and council is presented in Table 10.1.

A successful productivity and quality improvement program must have designated levels of responsibilities at various levels. In Chapter 1, the productivity and quality management hierarchy presented shows the various levels of emphasis. There has to be an organizational structure at the corporate, division, plant or site, and functional levels. At the corporate level, the vice president of productivity and quality management should be appointed to oversee all the division programs. Additional responsibility at this level includes project selection, review budget allocation to division programs, and corporate communications requirements for productivity and quality management. This level should have a close link to research and development, strategic planning, and business controls. The division directors should be appointed to oversee plant or site productivity and quality activities.

Table 10.1   Productivity and Quality Improvement Organizations

Organization	Responsibilities
Productivity steering committee	1. Provide leadership in monitoring organization policy to ensure that productivity and quality improvement goals and objectives are met 2. Provide leadership for initiating new programs 3. Review all projects and determine approval 4. Review resource allocation requirement by project and determine approval
Productivity council	1. Evaluate productivity and quality improvement project at the plant or functional level 2. Coordinate key good practice overall 3. Disseminate improvement ideas 4. Monitor activities related to productivity and quality, and recommend revision to policies where necessary
Productivity coordinators	1. Provide education and training 2. Provide effective communication internally and externally 3. Key contact for all issues 4. Design and implement new ideas 5. Adviser to operating units (departments) 6. Monitor factors affecting productivity and quality, and recommend corrective actions 7. Act as chair of the organization productivity and quality council 8. Initiate and maintain effective productivity and quality awareness programs

The role of the director is to ensure consistency and uniformity in the various plant locations. At this level, emphasis is placed on applied concepts, research, and techniques that improve productivity at the site level. The task force representing all plants are usually put in place to formulate overall productivity and quality improvement policies. The various plants usually appoint their own program managers or coordinators to act as a focal point on all key issues. The suggested

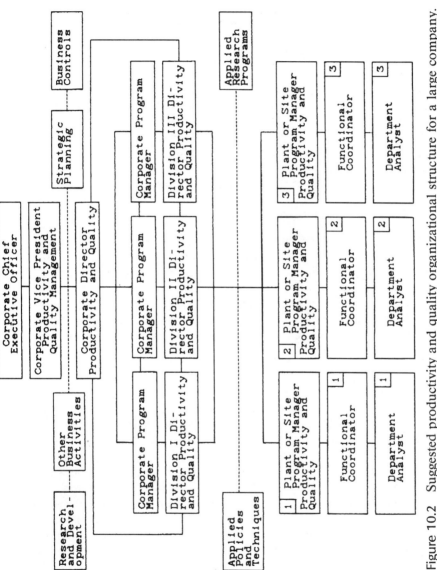

Figure 10.2 Suggested productivity and quality organizational structure for a large company.

organizational structure for a corporate productivity and quality program is presented in Figure 10.2. An organization may use their existing industrial engineering departments or functions as a focal point for coordinating productivity and quality improvement activities.

10.3 OVERCOMING PROBLEMS

Productivity and quality improvement must be promoted with enthusiasm and confidence. Everyone within the organization must be convinced that there is always a better way of improving the various tasks performed. However, many problems will arise that will present obstacles to retard the progress of the improvement program. The following common problems are usually experienced when promoting a productivity and quality improvement program.

10.3.1 Unwillingness to Change Work Habit

After learning a given pattern to perform a specific task, the pattern becomes a habit. It is often difficult to break away from the old habit and accept a new one. In addition to educating people about what the new productivity and quality improvement will bring to their tasks, people must be made to understand both the tangible and intangible benefits of changing from an old habit to a new one. A good strategy is to get everyone involved in all phases of the projects and programs when possible.

10.3.2 Resistance to Change

Anytime a new program or idea is promoted, people will accept or resist change depending on how they are affected by the change. There are four types of people one is likely to deal with when promoting productivity and quality improvements. There are those people who will do everything within their power to knock out the improvement because it creates an extra burden for them. Such people always find several reasons that the new improvement idea will not work. They always try to convince the improvement analyzer to stay with the old method of doing things. There are those people who do all the talking about how productivity and quality should be improved and promoted but take very little action. There are others who also spend their energy wishing someone else or some new system will automatically do the work. And finally there are those people with a keen interest in improving productivity and quality so that the organization can stay profitable. Such people get under the load, do the work, and provide all the support needed. The strategy is to be able to get all the four categories of people convinced to a great extent. This can be done through emphasis on teamwork and regular meetings to ensure effective communication of the selected improvement programs. It is also recommended that the cooperative efforts received be recognized among peers and supervisors.

10.3.3 Lack of Proper Planning

A productivity and quality improvement program can easily be unsuccessful if the requirements for start-up, key people, and total implementation of activities are not planned properly. It is important for the appropriate trained people with knowledge of productivity and quality be made coordinators of improvement projects. Such coordinators must specify in great detail the key activities needed for improvement projects, who is responsible, when they will be completed, and the expected benefit date.

10.3.4 Fear of the Unknown

Because new productivity and quality improvement ideas may have some uncertainty because of unproven results, most people are likely to be afraid of failure. As a result, they may resist the implementation of new concepts and ideas. All levels of management and employees must be educated about the essentials and importance of willingness to risk failure. If one new idea fails, it does not prevent another good one from succeeding.

10.3.5 Lack of Appropriate Data Base

A productivity and quality improvement program should be based on an adequate data base with correct historical information. It is every important to involve all levels of management and employees during improvement program start-ups to ensure that adequate data and a description of processes and procedures are provided.

10.3.6 Resentment or Criticism

Most people believe that they perform their task in the most effective and efficient manner and will resent criticism that proves otherwise. Criticism from productivity analysts, coordinators, and consultants should be offered constructively and positively to avoid backlash. On the other hand, those receiving ideas must be willing to do away with a defensive attitude. Everyone involved in the productivity and quality improvement program must have an open mind and the willingness to accommodate several different viewpoints.

10.3.7 Inadequate Sharing of Productivity and Quality Improvement Gains

Productivity and quality gains obtained from improvement programs must be distributed equitably and fairly to encourage a continued innovative process for additional ideas. Most organizations have a formal suggestion program, cost effectiveness program, bonus program, and awards program that are specifically designed to reward contributions to productivity and quality improvement. It is important to recognize that money is not the only source of motivation. Peer and superior recognition, promotions, additional challenges, job content, and others are also key sources of motivation.

10.3.8 Conflicting Compromise of Objectives

Productivity and quality improvement go together. Both managers and employees must stay away from the notion that if productivity improves, quality will suffer. The typical notion of push schedules when they are needed and push quality when it is wanted should basically be eliminated. Programs selected for improvements must satisfy dual objectives. People must be trained in how to manage conflicting objectives.

10.3.9 Complacency Resulting from Current Status

There is a tendency for those organizations that are already leaders in their respective industrial sector to feel so satisfied with their performance as leaders that they completely ignore the ongoing assessment of the method of operation and recommendations for implementable changes. Everyone within the organization must be trained to recognize that there is no limit to success through improvement. The more improvement ideas are implemented, the better off the organization will be financially.

10.3.10   Starting Off Too Big

Productivity and quality improvement issues can get complicated. The entire organization's problem cannot be resolved overnight. For improvement programs to be successful, they must be started with reasonable projects that the available resources can handle. Once the results and benefits of such projects have been obtained, bigger ventures can be considered.

10.3.11   Overall Strategy for Overcoming Common Problems

The following twelve points are recommended when promoting a productivity and quality improvement program:

1. Obtain the support of the total organization (management and employees).
2. Set realistic and opportunistic goals and objectives for productivity and quality improvement.
3. Plan program activities ahead of time.
4. Maintain internal and external contacts with key people who have the expertise to help and train workforce members in new techniques for productivity and quality improvement.
5. Establish a clear review process for concern and accomplishments that may arise.
6. Re-evaluate the program periodically, and make modifications when necessary.
7. Be flexible and willing to sacrifice time and other resources to obtain improvement.
8. Follow up continuously to resolve open issues.
9. Do not expect too great a productivity and quality improvement overnight.
10. Use expert opinions and advice when needed.
11. Obtain adequate staffing to complete the improvement program objectives.
12. Do not expect too great an immediate return on investment. Some improvement projects may have key benefits because they provide continuous growth in the long run.

Table 10.2   Checklist for Productivity and Quality Management

Self-assessment question for organization or task	Is question applicable? Yes	Is question applicable? No	Action plan to resolve weaknesses	Target date for completion	Benefit date for actions
1. Does the organization have a formal productivity and quality management program?					
2. How is the productivity and quality management organized? Is it effective?					
3. Does everyone within the organization have awareness of the emphasis on productivity and quality improvement?					
4. How is the productivity and quality management program at all levels?					
5. What educational programs are available to train employees and managers on the subject of productivity and quality?					
6. How effective are the training programs?					
7. Does the productivity and quality management have the support of senior executives?					
8. What are the avenues for implementing suggestions that affect productivity and quality?					
9. Do employees receive rewards for suggestions that improve productivity and quality?					
10. Are employees and managers paid for					

Table 10.2   (*Continued*)

Self-assessment question for organization or task	Is question applicable?		Action plan to resolve weaknesses	Target date for completion	Benefit date for actions
	Yes	No			
performance with consideration of overall improvement in technical and people productivity and quality?					
11. Does the organization have a comprehensive productivity and quality planning program?					
12. What are the strengths and weaknesses of the organization?					
13. What are the opportunities and threats facing the organization?					
14. What are the plan's central assumptions?					
15. What is the planning horizon (2, 5, or 10 years)?					
16. What key risks does the organization face?					
17. What is the probability of occurrence for each risk?					
18. What is the impact of each risk on resources and net economy?					
19. What are the strategies to capitalize on the strengths and opportunities?					
20. What are the key objectives for the future?					

(*continued*)

Table 10.2   (*Continued*)

Self-assessment question for organization or task	Is question applicable? Yes	No	Action plan to resolve weaknesses	Target date for completion	Benefit date for actions
21. What planning awareness programs are in place?					
22. What are the mechanisms for resource allocation?					
23. What are the mechanisms for plan evaluation?					
24. What are the mechanisms for plan maintenance?					
25. In what data base will the plan issues be tracked and maintained?					
26. What department, function, group of people, or individual is responsible for which planning task?					
27. What is the source of forecasted input variables?					
28. Who does the forecast and with what method?					
29. What is the organization's understanding of the economic trend?					
30. What is the organization understanding of legal or governmental influence? Geographical conditions? Social conditions?					
31. What is the strategy for market analysis and competition?					

Table 10.2 *(Continued)*

Self-assessment question for organization or task	Is question applicable? Yes	Is question applicable? No	Action plan to resolve weaknesses	Target date for completion	Benefit date for actions
Distribution channels?					
32. What are the communication flow patterns supporting the plans?					
33. What are the technology and production processes supporting the plan?					
34. What are other input variables supporting the plan (energy, capital, materials, labor, research and development)?					
35. What are the mechanisms for scheduling activities?					
36. What are the mechanisms for programming activities?					
37. What are the mechanisms for budgeting and controlling?					
38. What are implementation priorities based upon?					
39. What types of measures are in place to measure productivity and quality?					
40. What are the improvement strategies and techniques for productivity and quality?					

(continued)

Table 10.2   (*Continued*)

Self-assessment question for organization or task	Is question applicable? Yes	Is question applicable? No	Action plan to resolve weaknesses	Target date for completion	Benefit date for actions
41. How much research and development activity is there for productivity breakthroughs?					
42. Are the organizational structures balanced?					
43. Is the work force motivated to improve productivity and quality?					
44. Are mechanisms in place to eliminate job bottlenecks?					
45. Are too many design changes complicating the tasks?					
46. Are there too many inspection gates that delay the work flow?					
47. Is there a particular task that delays the entire flow of work?					
48. Is the teamwork concept practiced in resolving problems?					
49. Are there too many procedures to follow?					
50. Are there too many "nice to have" but not "needed" functions?					
51. Does the organization encourage healthy competition?					
52. Does the organization give incentives for technological innovations?					

Table 10.2   (*Continued*)

Self-assessment question for organization or task	Is question applicable? Yes	No	Action plan to resolve weaknesses	Target date for completion	Benefit date for actions
53. Are productivity and quality measures centered on only one area?					
54. Are there techniques for indirect efforts toward productivity control?					
55. Are there techniques for assessing quality in the business process?					
56. Is the work place design comfortable?					
57. Are tools designed properly for use by humans?					
58. Does the work place allow the smooth flow of work from one task to another?					
59. Are there safety hazards that impact on performance?					
60. Is there too much paperwork within the process?					
61. Are scrap costs too high?					
62. Are there too many nonvalue-adding operations?					
63. Does an employee handle the product more than once?					
64. Does the operator experience idle time?					
65. Are the methods for materials handling adequate?					

(*continued*)

Table 10.2   (*Continued*)

Self-assessment question for organization or task	Is question applicable? Yes	No	Action plan to resolve weaknesses	Target date for completion	Benefit date for actions
66. Are products and materials handled properly?					
67. Is the percentage of rejected work too high?					
68. Are defects coming from the supplier or vendor?					
69. Is the process inducing defects?					
70. Are there too many rework loops?					
71. Are the packaging materials adequate?					
72. Are employees matched with jobs for which they are best suited?					
73. Do the priorities for resources allocation follow the priorities of productivity and quality objectives?					
74. Are there enough communication sessions to discuss productivity and quality issues? Are there too many meetings, or too few?					
75. Are there mechanisms for balancing the work load among employees, departments, and functions?					
76. Are quality reports up-to-date?					
77. Do you have excessive overtime? Excessive absenteeism?					

Table 10.2 (*Continued*)

Self-assessment question for organization or task	Is question applicable?		Action plan to resolve weaknesses	Target date for completion	Benefit date for actions
	Yes	No			
78. Is the turnover rate of the work force high?					
79. Are there mechanisms for setting realistic standards and specifications?					
80. Does every employee understand the cost of waste?					
81. Are tools and equipment maintained properly?					
82. Do you have scheduled preventive maintenance?					
83. Is equipment availability time enough?					
84. Is equipment utilization time enough?					
85. Do you have employees who affect the productivity and quality of others through bad attitude?					
86. Do you have a mechanism for prioritizing jobs and assignments?					
87. Does management care about people's morale?					
88. Are people well paid for their performance?					
89. Do you have a difficult task that requires additional training, additional compensation?					

Table 10.2   (*Continued*)

Self-assessment question for organization or task	Is question applicable? Yes	Is question applicable? No	Action plan to resolve weaknesses	Target date for completion	Benefit date for actions
90. Do you have a job rotation plan to cross-train everyone?					
91. Is your work force stable?					
92. Are you hiring the right skill mix for jobs?					
93. Do you give awards for productivity and quality improvements?					
94. Do employees have adequate breaks, lunch, etc.?					
95. Are employees given allowance for task fatigue?					
96. Do employees have adequate development plans?					
97. Do employees have adequate performance plans?					
98. Is there equity in salary administration among employees?					
99. Is there a system for resolving grievances?					
100. Is there too little work specialization?					
101. Is there excessive work specialization?					
102. Are the same production or service problems recurring?					
103. Are there mechanisms in place to track ownership to problems?					

Table 10.2 (*Continued*)

Self-assessment question for organization or task	Is question applicable? Yes	No	Action plan to resolve weaknesses	Target date for completion	Benefit date for actions
104. Are improvements monitored regularly and reinforced?					
105. How is the executive time spent? On operational problems? On strategic problems? On new businesses?					
106. Are there management interferences across functions?					
107. Is there proper cross-communication across functions?					
108. Does everyone understand the organizational mission? Departmental mission? Goals and objectives?					
109. Are there production and service inconsistencies across different locations?					
110. Are the tasks too repetitious for employees?					
111. Do employees have basic controls over what they do?					
112. Are some employees performing tasks below the level of the skills for which they are being paid?					
113. Is technical creativity used properly?					

(*continued*)

Table 10.2 (*Continued*)

Self-assessment question for organization or task	Is question applicable? Yes	No	Action plan to resolve weaknesses	Target date for completion	Benefit date for actions
114. Are all the policies, procedures, and assignments top down?					
115. Is each activity and work element as specified worth doing?					
116. Does communication of work flow in the most simple, direct, and logical manner?					
117. Are there too many interruptions to the production or service caused by unplanned customer demand?					
118. Are there too many extra services requested by the customer?					
119. Does the organization have early manufacturing involvement in design phase?					
120. Is most time spent on important assignments or miscellaneous work?					
121. Is the quality standard for product and service too high?					
122. Are sampling inspections utilized in place of 100% inspection?					
123. Is work checked too many times at different processes and gates?					
124. Would quality errors be caught elsewhere if the present checks were eliminated?					

Table 10.2 (*Continued*)

Self-assessment question for organization or task	Is question applicable?		Action plan to resolve weaknesses	Target date for completion	Benefit date for actions
	Yes	No			
125. What would happen if the quality errors were not timely found?					
126. Are there cross-motions among people due to inadequate layout?					
127. Do documents flow smoothly from desk to desk?					
128. Can a computer be used to replace manual tracking efforts?					
129. Do individuals receiving memos, forms, and documents need them? Do they make use of them?					
130. What information is really necessary and useful to manage the task? The departments requirement's? Business required?					
131. Is the cost of data processing too high?					
132. Is the cost of materials too high? Can another vendor do the job?					
133. Is the cost of capital too high? Is an alternative financing method available?					
134. Is the cost of energy too high? What happens if solar energy is used? Oil? Gas? Water?					
135. Is the right equipment available when it is needed?					

(*continued*)

Table 10.2   (*Continued*)

Self-assessment question for organization or task	Is question applicable?		Action plan to resolve weaknesses	Target date for completion	Benefit date for actions
	Yes	No			
136. Can the organization afford a group technology concept?					
137. Does the organization have an adequate span of control?					
138. How much emphasis is placed on a management development program? Do managers practice a participative management approach?					
139. Are auditors used to evaluate processes and services?					
140. Are consultants used for new ideas?					
141. Are task forces used to resolve cross-functional problems?					
142. Are there measures for management and employee loyalty to the organization?					
143. Are there groups of employees working against organization tradition? Philosophy? Goals and objectives?					
144. Is there a balance between the existing system for employee reward and punishment?					
145. How are policy					

Table 10.2   (*Continued*)

Self-assessment question for organization or task	Is question applicable?		Action plan to resolve weaknesses	Target date for completion	Benefit date for actions
	Yes	No			
decisions formulated?					
146. Does the organization have an adequate executive development program?					
147. Are there executive talents that are wasted?					
148. What are the strategies for strong competition?					
149. What are the strategies for diversification of product and services?					
150. Is the marketing strategy adequate for the various product line and services offered?					
151. Do the organizational goals support long-term investment?					
152. How much is invested for future growth?					
153. Are there mechanisms in place to encourage technological innovation?					
154. What are the productivity and quality techniques needed as technology changes?					
155. Are there plans in place to manage a shorter cycle of technology?					
156. What is the organization's response rate to productivity and quality problems?					

10.4 CHECKLIST FOR PRODUCTIVITY AND QUALITY MANAGEMENT

It is often very easy to overlook issues or problems within the business process that are likely to contribute to overall productivity and quality improvement. It is also likely that overlooked problems may cause additional problems or have a negative impact on productivity and quality. This section focuses on providing a detailed checklist that will enable organizations to understand where problems are likely to occur. Some of the questions on the checklist (Table 10.2) may apply at the operational level; others may be useful at the management policy level. It is recommended that a key action plan be developed for areas of weaknesses. A specific task should be assigned to a specific function or person(s) with a target date for completion.

REFERENCES

Edosomwan, J. A., 1986a. A methodology for assessing the impact of robotics on total productivity in an assembly task. Proceedings of Annual International Industrial Engineering Conference, May, Dallas, Texas.

Edosomwan, J. A., 1986b. Productivity management in computer-aided manufacturing environment. Proceedings of First International Conference on Engineering Management, September 22–24, Washington, D.C.

Edosomwan, J. A., 1987. The meaning and management of productivity and quality. *Industrial Engineering*, January.

11

Conclusion

This book was written with a consideration of productivity and quality management as essential dimensions in an organization's effectiveness. It is important to note that productivity and quality are connected and interrelated. The goal-setting process, operations improvement, performance measurement, and other strategies geared toward organization improvement must be done with the understanding that productivity and quality are positively correlated. It is impossible to have one without the other. A sound productivity and quality management program can provide the basis for organizations to effectively address such issues as

Waste elimination within the organization

Elimination of process and procedural bottlenecks

Cost reduction

Defect reduction

Rapidly changing technology with increased sophistication and specialization

Rate of growth of diversified product lines and services

Other organizational complexities

In order for productivity and quality management efforts to continue to yield positive improvement and results, everyone within the organization must be willing to encourage teamwork. There must be cross-fertilization of ideas among people and across departments and functional areas. Everyone involved in the productivity and quality program must have the determination to improve the work organization effectively and efficiently. There must be strong awareness that competition can create a wide gap between two similar task, processes, firms, and profit levels obtained by similar organizations. Today, most companies in the United States and other countries worldwide are aware of the productivity and quality advantage enjoyed by Japanese companies. This is true in the electronics industry and automobile industry, and other areas as well. If the right things are going to be done regarding productivity and quality improvement, organizations must constantly train and retrain talents to provide the technical vitality needed to deal with pressing issues. Everyone within the organization must also believe that change is possible, change can bring prosperity, and change that brings improvement in organization should be welcome without caution and bottlenecks. Acceptance of changing values and challenges must also be done in conjunction with acceptance and implementation of new ideas, suggestions, and techniques.

Productivity and quality management is both a top-down and a bottom-up continuous process, top-down because senior management and executives must determine the right productivity and quality goals and objectives of each major entity within the organization. Management can set the operating climate, operating guidelines, and resource allocation that determine the successful implementation of productivity and quality improvement projects. Management can reject or take recommendations from subordinates in a manner that has a positive or negative impact on productivity and quality. In addition, top management establishes priorities for productivity and quality improvement projects based on the resources available. Productivity and quality management can be viewed as bottom-up because each employee responsible for each task or set of activities has the opportunity to evaluate their own mode of operations, eliminate the waste from such operations, design and develop better habits and methods of doing their own work, and implement actions on an ongoing basis to achieve improvements. It is important to realize that the skills requirements for productivity and quality improvement are not difficult to learn. The problem lies more with the continuous interference caused by old habits. Anyone willing to accept productivity and quality improvement challenges must be willing to break away from the old habit of doing things if it does not work well. One has to be willing to have an open mind for new ideas and to support their implementation. If everyone within the organization provides his or her support in implementing improvement ideas, the battle against inflation, high cost, and a poor quality product is won. It is also important to bring reality into the picture in business operations. Every productivity and quality issue is unique; no one skill or action is appropriate for all purposes. Management and employees should be trained to recognize the various basic distinctions to ensure progress in all given situations and in each area of the business.

The actions implemented to address productivity and quality issues and problems must be positive, consistent, and workable. In addition, management and employees must pay attention to business, technical, and people-oriented concerns on an on-going basis. Clearly, success is dependent upon the employees' and managers' belief in themselves, what they do, and how it is done. The success is also based on both the action plans and how they are applied. Measurement of productivity and quality improvement actions are necessary to see if expected results were obtained. The locus of control must focus attention on on-going continuous improvement and follow-up on open issues.

The book has provided significant tools, techniques, concepts, and ideas on how to improve productivity and quality in a highly competitive environment. The framework presented covers all aspects of business operations. Perhaps the materials provided has a broader use and applicability because it has been validated in real-life situations. I hope that the discussion throughout the book provides the understanding that productivity and quality are connected and require an ongoing process of continuous elimination of waste through the application of common sense. Productivity and quality improvement does not mean that people should work harder, but work smarter with better tools, techniques, processes, resources, and implementation of new ideas.

Appendix A

Statistical Tables*

Standard Normal Distribution
F-Distribution Table, $\alpha = 0.5$
F-Distribution Table, $\alpha = 0.1$
t-Distribution Table

---

*Adapted from *Process Control, Capability and Improvement*. The IBM Quality Institute Publication, Southbury, Connecticut, May 1985. Reprinted with permission.

Standard Normal Distribution
(Area Under Normal Curve)

a = the proportion of process output beyond a particular value of interest (such as a specification limit) that is z standard deviation units away from the process average (for a process that is in statistical control and is normally distributed). For example, if z = 2.17, a = .0150 or 1.5%. In any actual situation, this proportion is only approximate.

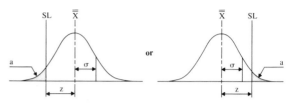

|z|	x.x0	x.x1	x.x2	x.x3	x.x4	x.x5	x.x6	x.x7	x.x8	x.x9
4.0	.00003									
3.9	.00005	.00005	.00004	.00004	.00004	.00004	.00004	.00004	.00003	.00003
3.8	.00007	.00007	.00007	.00006	.00006	.00006	.00006	.00005	.00005	.00005
3.7	.00011	.00010	.00010	.00010	.00009	.00009	.00008	.00008	.00008	.00008
3.6	.00016	.00015	.00015	.00014	.00014	.00013	.00013	.00012	.00012	.00011
3.5	.00023	.00022	.00022	.00021	.00020	.00019	.00019	.00018	.00017	.00017
3.4	.00034	.00032	.00031	.00030	.00029	.00028	.00027	.00026	.00025	.00024
3.3	.00048	.00047	.00045	.00043	.00042	.00040	.00039	.00038	.00036	.00035
3.2	.00069	.00066	.00064	.00062	.00060	.00058	.00056	.00054	.00052	.00050
3.1	.00097	.00094	.00090	.00087	.00084	.00082	.00079	.00076	.00074	.00071
3.0	.00135	.00131	.00126	.00122	.00118	.00114	.00111	.00107	.00104	.00100
2.9	.0019	.0018	.0018	.0017	.0016	.0016	.0015	.0015	.0014	.0014
2.8	.0026	.0025	.0024	.0023	.0023	.0022	.0021	.0021	.0020	.0019
2.7	.0035	.0034	.0033	.0032	.0031	.0030	.0029	.0028	.0027	.0026
2.6	.0047	.0045	.0044	.0043	.0041	.0040	.0039	.0038	.0037	.0036
2.5	.0062	.0060	.0059	.0057	.0055	.0054	.0052	.0051	.0049	.0048
2.4	.0082	.0080	.0078	.0075	.0073	.0071	.0069	.0068	.0066	.0064
2.3	.0107	.0104	.0102	.0099	.0096	.0094	.0091	.0089	.0087	.0084
2.2	.0139	.0136	.0132	.0129	.0125	.0122	.0119	.0116	.0113	.0110
2.1	.0179	.0174	.0170	.0166	.0162	.0158	.0154	.0150	.0146	.0143
2.0	.0228	.0222	.0217	.0212	.0207	.0202	.0197	.0192	.0188	.0183
1.9	.0287	.0281	.0274	.0268	.0262	.0256	.0250	.0244	.0239	.0233
1.8	.0359	.0351	.0344	.0336	.0329	.0322	.0314	.0307	.0301	.0294
1.7	.0446	.0436	.0427	.0418	.0409	.0401	.0392	.0384	.0375	.0367
1.6	.0548	.0537	.0526	.0516	.0505	.0495	.0485	.0475	.0465	.0455
1.5	.0668	.0655	.0643	.0630	.0618	.0606	.0594	.0582	.0571	.0559
1.4	.0808	.0793	.0778	.0764	.0749	.0735	.0721	.0708	.0694	.0681
1.3	.0968	.0951	.0934	.0918	.0901	.0885	.0869	.0853	.0838	.0823
1.2	.1151	.1131	.1112	.1093	.1075	.1056	.1038	.1020	.1003	.0985
1.1	.1357	.1335	.1314	.1292	.1271	.1251	.1230	.1210	.1190	.1170
1.0	.1587	.1562	.1539	.1515	.1492	.1469	.1446	.1423	.1401	.1379
0.9	.1841	.1814	.1788	.1762	.1736	.1711	.1685	.1660	.1635	.1611
0.8	.2119	.2090	.2061	.2033	.2005	.1977	.1949	.1922	.1894	.1867
0.7	.2420	.2389	.2358	.2327	.2297	.2266	.2236	.2206	.2177	.2148
0.6	.2743	.2709	.2676	.2643	.2611	.2578	.2546	.2514	.2483	.2451
0.5	.3085	.3050	.3015	.2981	.2946	.2912	.2877	.2843	.2810	.2776
0.4	.3446	.3409	.3372	.3336	.3300	.3264	.3228	.3192	.3156	.3121
0.3	.3821	.3783	.3745	.3707	.3669	.3632	.3594	.3557	.3520	.3483
0.2	.4207	.4168	.4129	.4090	.4052	.4013	.3974	.3936	.3897	.3859
0.1	.4602	.4562	.4522	.4483	.4443	.4404	.4364	.4325	.4286	.4247
0.0	.5000	.4960	.4920	.4880	.4840	.4801	.4761	.4721	.4681	.4641

F-Distribution Table, $\alpha = 0.5$

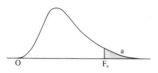

d.f.2 (denom-inator)	d.f.1 (numerator)							
	1	2	3	4	5	6	7	8
1	161	200	216	225	230	234	237	239
2	18.5	19.0	19.2	19.2	19.3	19.3	19.4	19.4
3	10.1	9.55	9.28	9.12	9.01	8.94	8.89	8.85
4	7.71	6.94	6.59	6.39	6.26	6.16	6.09	6.04
5	6.61	5.79	5.41	5.19	5.05	4.95	4.88	4.82
6	5.99	5.14	4.76	4.53	4.39	4.28	4.21	4.15
7	5.59	4.74	4.35	4.12	3.97	3.87	3.79	3.73
8	5.32	4.46	4.07	3.84	3.69	3.58	3.50	3.44
9	5.12	4.26	3.86	3.63	3.48	3.37	3.29	3.23
10	4.96	4.10	3.71	3.48	3.33	3.22	3.14	3.07
11	4.84	3.98	3.59	3.36	3.20	3.09	3.01	2.95
12	4.75	3.89	3.49	3.26	3.11	3.00	2.91	2.85
13	4.67	3.81	3.41	3.18	3.03	2.92	2.83	2.77
14	4.60	3.74	3.34	3.11	2.96	2.85	2.76	2.70
15	4.54	3.68	3.29	3.06	2.90	2.79	2.71	2.64
16	4.49	3.63	3.24	3.01	2.85	2.74	2.66	2.59
17	4.45	3.59	3.20	2.96	2.81	2.70	2.61	2.55
18	4.41	3.55	3.16	2.93	2.77	2.66	2.58	2.51
19	4.38	3.52	3.13	2.90	2.74	2.63	2.54	2.48
20	4.35	3.49	3.10	2.87	2.71	2.60	2.51	2.45
21	4.32	3.47	3.07	2.84	2.68	2.57	2.49	2.42
22	4.30	3.44	3.05	2.82	2.66	2.55	2.46	2.40
23	4.28	3.42	3.03	2.80	2.64	2.53	2.44	2.37
24	4.26	3.40	3.01	2.78	2.62	2.51	2.42	2.36
25	4.24	3.39	2.99	2.76	2.60	2.49	2.40	2.34
30	4.17	3.32	2.92	2.69	2.53	2.42	2.33	2.27
40	4.08	3.23	2.84	2.61	2.45	2.34	2.25	2.18
60	4.00	3.15	2.76	2.53	2.37	2.25	2.17	2.10
120	3.92	3.07	2.68	2.45	2.29	2.18	2.09	2.02
∞	3.84	3.00	2.60	2.37	2.21	2.10	2.01	1.94

					d.f.1 (numerator)					
9	10	12	15	20	24	30	40	60	120	∞
241	242	244	246	248	249	250	251	252	253	254
19.4	19.4	19.4	19.4	19.4	19.5	19.5	19.5	19.5	19.5	19.5
8.81	8.79	8.74	8.70	8.66	8.64	8.62	8.59	8.57	8.55	8.53
6.00	5.96	5.91	5.86	5.80	5.77	5.75	5.72	5.69	5.66	5.63
4.77	4.74	4.68	4.62	4.56	4.53	4.50	4.46	4.43	4.40	4.37
4.10	4.06	4.00	3.94	3.87	3.84	3.81	3.77	3.74	3.70	3.67
3.68	3.64	3.57	3.51	3.44	3.41	3.38	3.34	3.30	3.27	3.23
3.39	3.35	3.28	3.22	3.15	3.12	3.08	3.04	3.01	2.97	2.93
3.18	3.14	3.07	3.01	2.94	2.90	2.86	2.83	2.79	2.75	2.71
3.02	2.98	2.91	2.85	2.77	2.74	2.70	2.66	2.62	2.58	2.54
2.90	2.85	2.79	2.72	2.65	2.61	2.57	2.53	2.49	2.45	2.40
2.80	2.75	2.69	2.62	2.54	2.51	2.47	2.43	2.38	2.34	2.30
2.71	2.67	2.60	2.53	2.46	2.42	2.38	2.34	2.30	2.25	2.21
2.65	2.60	2.53	2.46	2.39	2.35	2.31	2.27	2.22	2.18	2.13
2.59	2.54	2.48	2.40	2.33	2.29	2.25	2.20	2.16	2.11	2.07
2.54	2.49	2.42	2.35	2.28	2.24	2.19	2.15	2.11	2.06	2.01
2.49	2.45	2.38	2.31	2.23	2.19	2.15	2.10	2.06	2.01	1.96
2.46	2.41	2.34	2.27	2.19	2.15	2.11	2.06	2.02	1.97	1.92
2.42	2.38	2.31	2.23	2.16	2.11	2.07	2.03	1.98	1.93	1.88
2.39	2.35	2.28	2.20	2.12	2.08	2.04	1.99	1.95	1.90	1.84
2.37	2.32	2.25	2.18	2.10	2.05	2.01	1.96	1.92	1.87	1.81
2.34	2.30	2.23	2.15	2.07	2.03	1.98	1.94	1.89	1.84	1.78
2.32	2.27	2.20	2.13	2.05	2.01	1.96	1.91	1.86	1.81	1.76
2.30	2.25	2.18	2.11	2.03	1.98	1.94	1.89	1.84	1.79	1.73
2.28	2.24	2.16	2.09	2.01	1.96	1.92	1.87	1.82	1.77	1.71
2.21	2.16	2.09	2.01	1.93	1.89	1.84	1.79	1.74	1.68	1.62
2.12	2.08	2.00	1.92	1.84	1.79	1.74	1.69	1.64	1.58	1.51
2.04	1.99	1.92	1.84	1.75	1.70	1.65	1.59	1.53	1.47	1.39
1.96	1.91	1.83	1.75	1.66	1.61	1.55	1.50	1.43	1.35	1.25
1.88	1.83	1.75	1.67	1.57	1.52	1.46	1.39	1.32	1.22	1.00

F-Distribution Table, $\alpha = .01$

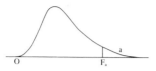

d.f.2 (denominator)	d.f.1 (numerator)							
	1	2	3	4	5	6	7	8
1	4,052	5,000	5,403	5,625	5,764	5,859	5,928	5,982
2	98.5	99.0	99.2	99.2	99.3	99.3	99.4	99.4
3	34.1	30.8	29.5	28.7	28.2	27.9	27.7	27.5
4	21.2	18.0	16.7	16.0	15.5	15.2	15.0	14.8
5	16.3	13.3	12.1	11.4	11.0	10.7	10.5	10.3
6	13.7	10.9	9.78	9.15	8.75	8.47	8.26	8.10
7	12.2	9.55	8.45	7.85	7.46	7.19	6.99	6.84
8	11.3	8.65	7.59	7.01	6.63	6.37	6.18	6.03
9	10.6	8.02	6.99	6.42	6.06	5.80	5.61	5.47
10	10.0	7.56	6.55	5.99	5.64	5.39	5.20	5.06
11	9.65	7.21	6.22	5.67	5.32	5.07	4.89	4.74
12	9.33	6.93	5.95	5.41	5.06	4.82	4.64	4.50
13	9.07	6.70	5.74	5.21	4.86	4.62	4.44	4.30
14	8.86	6.51	5.56	5.04	4.70	4.46	4.28	4.14
15	8.68	6.36	5.42	4.89	4.56	4.32	4.14	4.00
16	8.53	6.23	5.29	4.77	4.44	4.20	4.03	3.89
17	8.40	6.11	5.19	4.67	4.34	4.10	3.93	3.79
18	8.29	6.01	5.09	4.58	4.25	4.01	3.84	3.71
19	8.19	5.93	5.01	4.50	4.17	3.94	3.77	3.63
20	8.10	5.85	4.94	4.43	4.10	3.87	3.70	3.56
21	8.02	5.78	4.87	4.37	4.04	3.81	3.64	3.51
22	7.95	5.72	4.82	4.31	3.99	3.76	3.59	3.45
23	7.88	5.66	4.76	4.26	3.94	3.71	3.54	3.41
24	7.82	5.61	4.72	4.22	3.90	3.67	3.50	3.36
25	7.77	5.57	4.68	4.18	3.86	3.63	3.46	3.32
30	7.56	5.39	4.51	4.02	3.70	3.47	3.30	3.17
40	7.31	5.18	4.31	3.83	3.51	3.29	3.12	2.99
60	7.08	4.98	4.13	3.65	3.34	3.12	2.95	2.82
120	6.85	4.79	3.95	3.48	3.17	2.96	2.79	2.66
∞	6.63	4.61	3.78	3.32	3.02	2.80	2.64	2.51

				d.f.1. (numerator)						
9	10	12	15	20	24	30	40	60	120	∞
6,023	6,056	6,106	6,157	6,209	6,235	6,261	6,287	6,313	6,339	6,366
99.4	99.4	99.4	99.4	99.4	99.5	99.5	99.5	99.5	99.5	99.5
27.3	27.2	27.1	26.9	26.7	26.6	26.5	26.4	26.3	26.2	26.1
14.7	14.5	14.4	14.2	14.0	13.9	13.8	13.7	13.7	13.6	13.5
10.2	10.1	9.89	9.72	9.55	9.47	9.38	9.29	9.20	9.11	9.02
7.98	7.87	7.72	7.56	7.40	7.31	7.23	7.14	7.06	6.97	6.88
6.72	6.62	6.47	6.31	6.16	6.07	5.99	5.91	5.82	5.74	5.65
5.91	5.81	5.67	5.52	5.36	5.28	5.20	5.12	5.03	4.95	4.86
5.35	5.26	5.11	4.96	4.81	4.73	4.65	4.57	4.48	4.40	4.31
4.94	4.85	4.71	4.56	4.41	4.33	4.25	4.17	4.08	4.00	3.91
4.63	4.54	4.40	4.25	4.10	4.02	3.94	3.86	3.78	3.69	3.60
4.39	4.30	4.16	4.01	3.86	3.78	3.70	3.62	3.54	3.45	3.36
4.19	4.10	3.96	3.82	3.66	3.59	3.51	3.43	3.34	3.25	3.17
4.03	3.94	3.80	3.66	3.51	3.43	3.35	3.27	3.18	3.09	3.00
3.89	3.80	3.67	3.52	3.37	3.29	3.21	3.13	3.05	2.96	2.87
3.78	3.69	3.55	3.41	3.26	3.18	3.10	3.02	2.93	2.84	2.75
3.68	3.59	3.46	3.31	3.16	3.08	3.00	2.92	2.83	2.75	2.65
3.60	3.51	3.37	3.23	3.08	3.00	2.92	2.84	2.75	2.66	2.57
3.52	3.43	3.30	3.15	3.00	2.92	2.84	2.76	2.67	2.58	2.49
3.46	3.37	3.23	3.09	2.94	2.86	2.78	2.69	2.61	2.52	2.42
3.40	3.31	3.17	3.03	2.88	2.80	2.72	2.64	2.55	2.46	2.36
3.35	3.26	3.12	2.98	2.83	2.75	2.67	2.58	2.50	2.40	2.31
3.30	3.21	3.07	2.93	2.78	2.70	2.62	2.54	2.45	2.35	2.26
3.26	3.17	3.03	2.89	2.74	2.66	2.58	2.49	2.40	2.31	2.21
3.22	3.13	2.99	2.85	2.70	2.62	2.53	2.45	2.36	2.27	2.17
3.07	2.98	2.84	2.70	2.55	2.47	2.39	2.30	2.21	2.11	2.01
2.89	2.80	2.66	2.52	2.37	2.29	2.20	2.11	2.02	1.92	1.80
2.72	2.63	2.50	2.35	2.20	2.12	2.03	1.94	1.84	1.73	1.60
2.56	2.47	2.34	2.19	2.03	1.95	1.86	1.76	1.66	1.53	1.38
2.41	2.32	2.18	2.04	1.88	1.79	1.70	1.59	1.47	1.32	1.00

t-Distribution Table

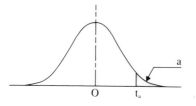

d.f.	$t_{.100}$	$t_{.050}$	$t_{.025}$	$t_{.010}$	$t_{.005}$
1	3.078	6.314	12.706	31.821	63.657
2	1.886	2.920	4.303	6.965	9.925
3	1.638	2.353	3.182	4.541	5.841
4	1.533	2.132	2.776	3.747	4.604
5	1.476	2.015	2.571	3.365	4.032
6	1.440	1.943	2.447	3.143	3.707
7	1.415	1.895	2.365	2.998	3.499
8	1.397	1.860	2.306	2.896	3.355
9	1.383	1.833	2.262	2.821	3.250
10	1.372	1.812	2.228	2.764	3.169
11	1.363	1.796	2.201	2.718	3.106
12	1.356	1.782	2.179	2.681	3.055
13	1.350	1.771	2.160	2.650	3.012
14	1.345	1.761	2.145	2.624	2.977
15	1.341	1.753	2.131	2.602	2.947
16	1.337	1.746	2.120	2.583	2.921
17	1.333	1.740	2.110	2.567	2.898
18	1.330	1.734	2.101	2.552	2.878
19	1.328	1.729	2.093	2.539	2.861
20	1.325	1.725	2.086	2.528	2.845
21	1.323	1.721	2.080	2.518	2.831
22	1.321	1.717	2.074	2.508	2.819
23	1.319	1.714	2.069	2.500	2.807
24	1.318	1.711	2.064	2.492	2.797
25	1.316	1.708	2.060	2.485	2.787
26	1.315	1.706	2.056	2.479	2.779
27	1.314	1.703	2.052	2.473	2.771
28	1.313	1.701	2.048	2.467	2.763
29	1.311	1.699	2.045	2.462	2.756
∞	1.282	1.645	1.960	2.326	2.576

Appendix B

Blank Work Forms for Statistical Process Control Implementation

---

*Adapted from *Process Control, Capability and Improvement*. The IBM Quality Institute Publication, Southbury, Connecticut, May 1985. Reprinted with permission.

Statistical Process Control Histogram

Part/Asm. Name	Machine	Operation Type:	Nominal
Part No.	Brass Tag No.	☐ Sub. Asm.	Upper Spec. Limit
Parameter	Tool No.	☐ Sub. Asm. Test	Lower Spec. Limit
Operation	Start Date	☐ Final Asm.	Gaging Method
Department	End Date	☐ Final Test	Unit of Measure
		☐ Other	Sample Frequency

Dim.

5   10   15   20   25   30   35   40   45   No.   %   Cum. %

Total

Operators

Sketch of Dimension

Normal Probability Scale

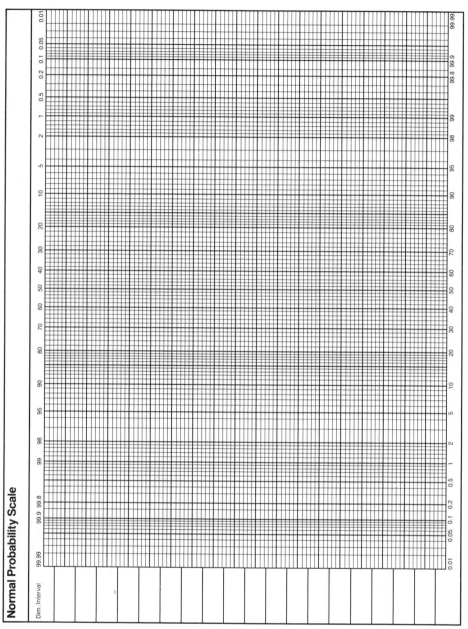

Dim. Interval

Variables Control Chart X̄&R

Averages & Ranges

Part/Asm. Name	Operation	Specification	Chart No.
Part No.	Department	Gage	Unit of Measure
Parameter	Machine	Sample Size Frequency	Zero Equals

Date				
Time				
Operator				
Sample Measurements	1			
	2			
	3			
	4			
	5			
Sum				
Average X̄				
Range R				

Averages

Ranges

Variables Control Chart $\overline{X}$&R   Averages & Ranges
Calculation Worksheet

Preliminary Control Limits (based on subgroups: _____)	Revised Control Limits (if necessary)

Calculate Average Range:

$\overline{R} = \dfrac{\Sigma R}{k} =$ _____ $=$ _____

$\overline{R} =$ _____ $=$ _____

Σ = Sum of, and k = number of subgroups

Calculate Control Limits for Ranges:

$UCL_R = D_4 \times \overline{R} =$ _____ $\times$ _____ $=$ _____

$UCL_R =$ _____ $\times$ _____ $=$ _____

$LCL_R = D_3 \times \overline{R} =$ _____ $\times$ _____ $=$ _____

$LCL_R =$ _____ $\times$ _____ $=$ _____

Calculate Grand Average:

$\overline{\overline{X}} = \dfrac{\Sigma \overline{X}}{k} =$ _____ $=$ _____

$\overline{\overline{X}} =$ _____ $=$ _____

Calculate Control Limits for Averages:

$A_2\overline{R} =$ _____ $\times$ _____ $=$ _____

$A_2\overline{R} =$ _____ $\times$ _____ $=$ _____

$UCL_{\overline{x}} = \overline{\overline{X}} + A_2\overline{R} =$ _____ $+$ _____ $=$ _____

$UCL_{\overline{x}} =$ _____ $+$ _____ $=$ _____

$LCL_{\overline{x}} = \overline{\overline{X}} - A_2\overline{R} =$ _____ $-$ _____ $=$ _____

$LCL_{\overline{x}} =$ _____ $-$ _____ $=$ _____

Estimate of Standard Deviation (If the process is in statistical control):

$\sigma = \dfrac{\overline{R}}{d_2} =$ _____

Subgroup Size n	A_2	D_3	D_4	d_2	Subgroup Size n	A_2	D_3	D_4	d_2
2	1.880	*	3.267	1.128	6	0.483	*	2.004	2.534
3	1.023	*	2.575	1.693	7	0.419	0.076	1.924	2.704
4	0.729	*	2.282	2.059	8	0.373	0.136	1.864	2.847
5	0.577	*	2.115	2.326	9	0.337	0.184	1.816	2.970
					10	0.308	0.223	1.777	3.078

*Lower control limits for R do not exist for sample sizes below 7.

Notes: Record all adjustments, tool changes, etc.			
Subgroup No.	Date	Time	Comments

Variables Control Chart X̄&S   Averages & Sample Standard Deviations

Part/Asm. Name	Operation	Specification	Chart No.
Part No.	Department	Gage	Unit of Measure
Parameter	Machine	Sampling Frequency	Zero Equals

Date																								
Time																								
Operator																								
Sample Size																								
Average X̄																								
Std. Dev. S																								

Averages

Standard Deviations

Variables Control Chart X̄&S   Averages & Sample Standard Deviations
Calculation Worksheet

Calculate Average Standard Deviation:

$\bar{S} = \dfrac{\Sigma S}{k} =$ _____ = _____

where Σ = sum of, and k = number of subgroups

Calculate Control Limits for Standard Deviation:

$UCL_S = B_4 \times \bar{S} =$ _____ x _____ = _____

$LCL_S = B_3 \times \bar{S} =$ _____ x _____ = _____

Calculate Grand Averages:

$\bar{\bar{X}} = \dfrac{\Sigma \bar{X}}{k} =$ _____ = _____

Calculate Control Limits for Averages:

$A_3\bar{S} =$ _____ x _____ = _____

$UCL_{\bar{x}} = \bar{\bar{X}} + A_3\bar{S} =$ _____ + _____ = _____

$LCL_{\bar{x}} = \bar{\bar{X}} - A_3\bar{S} =$ _____ − _____ = _____

Estimate of Process Standard Deviation:
(If the process is in statistical control)

$\hat{\sigma} = \bar{S}_{/C_4} =$ _____ = _____

Subgroup Size					Subgroup Size				
n	A_3	B_3	B_4	c_4	n	A_3	B_3	B_4	c_4
2	2.659	*	3.267	0.7979	16	0.763	0.448	1.552	0.9835
3	1.954	*	2.568	0.8862	17	0.739	0.466	1.534	0.9845
4	1.628	*	2.266	0.9213	18	0.718	0.482	1.518	0.9854
5	1.427	*	2.089	0.9400	19	0.698	0.497	1.503	0.9862
					20	0.680	0.510	1.490	0.9869
6	1.287	0.030	1.970	0.9515					
7	1.182	0.118	1.882	0.9594	21	0.663	0.523	1.477	0.9876
8	1.099	0.185	1.815	0.9650	22	0.647	0.534	1.466	0.9882
9	1.032	0.239	1.761	0.9693	23	0.633	0.545	1.455	0.9887
10	0.975	0.284	1.716	0.9727	24	0.619	0.555	1.445	0.9892
					25	0.606	0.565	1.435	0.9896
11	0.927	0.321	1.679	0.9754					
12	0.886	0.354	1.646	0.9776					
13	0.850	0.382	1.618	0.9794					
14	0.817	0.406	1.594	0.9810					
15	0.789	0.428	1.572	0.9823					

*There is no lower control limit for standard deviations for sample sizes below 6.

Notes: Record all adjustments, tool changes, etc.

Subgroup No.	Date	Time	Comments

Variables Control Chart for Medians X̃&R Medians & Ranges

Part/Asm. Name	Operation	Specification	Chart No.
Part No.	Department	Gage	Unit of Measure
Parameter	Machine	Sample Size/Frequency	Zero Equals

	Date
	Time
	Operator
Sample Measurements	1
	2
	3
	4
	5
	Sum
	Median X̃
	Range R

Medians

Ranges

1 2 3 4 5 6 7 8 9 10 11 12 13 14 15 16 17 18 19 20 21 22 23 24 25

Variables Control Chart for Medians $\tilde{X}$&R Medians & Ranges
Calculation Worksheet

Preliminary Control Limits (based on subgroups: _____) | **Revised Control Limits** (if necessary)

Calculate Average Range:

$\bar{R} = \dfrac{\Sigma R}{k}$ = _____ = _____

Σ = Sum of, and k = number of subgroups

$\bar{R}$ = _____ = _____

Calculate Control Limits for Ranges:

$UCL_R = D_4 \times \bar{R}$ = _____ x _____ = _____

$LCL_R = D_3 \times \bar{R}$ = _____ x _____ = _____

UCL_R = _____ x _____ = _____

LCL_R = _____ x _____ = _____

Calculate Average of Medians:

$\bar{\tilde{X}} = \dfrac{\Sigma \tilde{X}}{k}$ = _____ = _____

$\bar{\tilde{X}}$ = _____ = _____

Calculate Control Limits for Medians:

$\tilde{A}_2 \bar{R}$ = _____ x _____ = _____

$UCL_{\tilde{X}} = \bar{\tilde{X}} + \tilde{A}_2 \bar{R}$ = _____ + _____ = _____

$LCL_{\tilde{X}} = \bar{\tilde{X}} - \tilde{A}_2 \bar{R}$ = _____ - _____ = _____

$\tilde{A}_2 \bar{R}$ = _____ x _____ = _____

$UCL_{\tilde{X}}$ = _____ + _____ = _____

$LCL_{\tilde{X}}$ = _____ - _____ = _____

Estimate of Standard Deviation (If the process is in statistical control):

$\sigma = \dfrac{\bar{R}}{d_2}$ = _____

Subgroup Size					Subgroup Size				
n	$\tilde{A}_2$	D_3	D_4	d_2	n	$\tilde{A}_2$	D_3	D_4	d_2
2	1.88	*	3.267	1.128	6	0.55	*	2.004	2.534
3	1.19	*	2.575	1.693	7	0.51	0.076	1.924	2.704
4	0.80	*	2.282	2.059	8	0.43	0.136	1.864	2.847
5	0.69	*	2.115	2.326	9	0.41	0.184	1.816	2.970
					10	0.36	0.223	1.777	3.078

*Lower control limits for R do not exist for sample sizes below 7.

Notes: Record all adjustments, tool changes, etc.

Subgroup No.	Date	Time	Comments

Control Chart for Attribute Data

Part Asm. Name	Operation	Nonconforming Units:	Nonconformities:	Chart No.
Part No.	Department	☐ np	☐ c	Average Sample Size
Parameter	Specification	☐ p	☐ u	Frequency

	Sample Size (n)
Discrep-ancies	Number (np.c)
	Proportion (p.u)
	Date

1 2 3 4 5 6 7 8 9 10 11 12 13 14 15 16 17 18 19 20 21 22 23 24 25

Notes: Record all adjustments, tool changes, etc.

Subgroup No.	Date	Time	Comments

Any change in people, materials, equipment, methods or environment should be noted. These notes will help you to take corrective or process improvement action when signaled by the control chart.

Continued on back

Control Chart for Attribute Data
Calculation Worksheet

	Nonconforming Units	**Nonconformities**

Number

(Subgroup sizes must be equal.)

<u>np Chart</u>

$$\text{UCL } np, \text{ LCL } np = n\bar{p} \pm 3\sqrt{n\bar{p}\left(1 - \frac{n\bar{p}}{n}\right)}$$

<u>c Chart</u>

$$\text{UCL } c, \text{ LCL } c = \bar{c} \pm 3\sqrt{\bar{c}}$$

Proportion

(Subgroup sizes need not be equal.)

<u>p Chart</u>

$$\text{UCL } p, \text{ LCL } p = \bar{p} \pm 3\sqrt{\frac{\bar{p}(1\cdot\bar{p})}{n}}$$

<u>u Chart</u>

$$\text{UCL } u, \text{ LCL } u = \bar{u} \pm 3\sqrt{\frac{\bar{u}}{n}}$$

$$\text{or} = \bar{p} \pm \frac{3\sqrt{\bar{p}(1-\bar{p})}}{\sqrt{n}}$$

$$\text{or} = \bar{u} \pm \frac{3\sqrt{\bar{u}}}{\sqrt{n}}$$

Notes: Record all adjustments, tool changes, etc.

Subgroup No.	Date	Time	Comments

Variables Control Chart for Individuals X&R Individuals & Ranges

Part.Asm. Name	Operation	Specification	Chart No.
Part No.	Department	Gage	Unit of Measure
Parameter	Machine	Sample Size Frequency	Zero Equals

Date																								
Time																								
Operator																								
Sum																								
Average X̄																								
Moving Range																								

Individuals

Moving Ranges

Variables Control Chart for Individuals X&R Individuals & Ranges
Calculation Worksheet

Preliminary Control Limits (based on subgroups: _____)   **Revised Control Limits** (if necessary)

Calculate Average Range:

$\bar{R} = \dfrac{\Sigma R}{k} =$ _____ = _____ $\bar{R} =$ _____ = _____

Σ = Sum of, and k = number of subgroups

Calculate Control Limits for Ranges:

$UCL_R = D_4 \times \bar{R} =$ _____ x _____ = _____ $UCL_{\bar{R}} =$ _____ x _____ = _____

$LCL_R = D_3 \times \bar{R} =$ _____ x _____ = _____ $LCL_{\bar{R}} =$ _____ x _____ = _____

Calculate Process Average:

$\bar{X} = \dfrac{\Sigma X}{k} =$ _____ = _____ $\bar{X} =$ _____ = _____

Calculate Control Limits for Individuals:

$E_2\bar{R} =$ _____ x _____ = _____ $E_2\bar{R} =$ _____ x _____ = _____

$UCL_x = \bar{X} + E_2\bar{R} =$ _____ + _____ = _____ $UCL_x =$ _____ + _____ = _____

$LCL_x = \bar{X} - E_2\bar{R} =$ _____ - _____ = _____ $LCL_x =$ _____ - _____ = _____

Estimate of Standard Deviation (If the process is in statistical control):

$\sigma = \dfrac{\bar{R}}{d_2} =$ _____

Subgroup Size					Subgroup Size				
n	E_2	D_3	D_4	d_2	n	E_2	D_3	D_4	d_2
2	2.660	*	3.267	1.128	6	1.184	*	2.004	2.534
3	1.772	*	2.575	1.693	7	1.109	0.076	1.924	2.704
4	1.457	*	2.282	2.059	8	1.054	0.136	1.864	2.847
5	1.290	*	2.115	2.326	9	1.010	0.184	1.816	2.970
					10	0.975	0.223	1.777	3.078

*Lower control limits for R do not exist for sample sizes below 7.

Notes: Record all adjustments, tool changes, etc.

Subgroup No.	Date	Time	Comments

Measurement System/Gage Capability Calculation Worksheet

Part/Asm. Name	Gage Name	Part No.
Characteristic	Gage No.	Measurement Unit
Specification	Gage Type	Zero Equals

Operator	A				B				C			
Sample No.	1st Trial	2nd Trial	3rd Trial	Range	1st Trial	2nd Trial	3rd Trial	Range	1st Trial	2nd Trial	3rd Trial	Range
1												
2												
3												
4												
5												
6												
7												
8												
9												
10												
Totals												

$\bar{R}_A =$

$\bar{R}_B =$

$\bar{R}_C =$

Sum

$\bar{X}_A$

Sum

$\bar{X}_B$

Sum

$\bar{X}_C$

Test for Control:

Upper Control Limit, $UCL_R = D_4\bar{R} =$ _____ x _____ = _____

where: $\bar{R}$ is the average of: $\bar{R}_A + \bar{R}_B + \bar{R}_C =$ _____ + _____ + _____ = _____

$D_4 = 3.27$ for 2 trials or 2.58 for 3 trials.

If any individual range exceeds this limit, the measurement or reading should be reviewed, repeated, corrected or discarded as appropriate, and new averages and ranges should be computed.

Measurement System/Gage Capability Analysis:

Equipment Variation ("Repeatability") $= K_1\bar{R} =$ _____ x _____ = [ ] Repeatability

where: $K_1 = 4.56$ for 2 trials or 3.05 for 3 trials.

Operator Variation ("Reproducibility") $= K_2\bar{X}_{diff} =$ _____ x _____ = [ ] Reproducibility

where: $K_2 = 3.65$ for 2 operators or 2.70 for 3 operators.

$\bar{X}_{diff}$ is the difference between the $max_{\bar{x}}$ and $min_{\bar{x}}$.

Total "Repeatability" and "Reproducibility" Variation (R&R) $\sqrt{(\text{Repeatability})^2 + (\text{Reproducibility})^2}$ Gage Capability

$= \sqrt{(\underline{\quad})^2 + (\underline{\quad})^2} = \sqrt{\underline{\quad} + \underline{\quad}} = \sqrt{\underline{\quad}} =$ [ ]

Gage Acceptability Determination:

$\dfrac{\text{Total Gage Capability (R\&R)}}{\text{Specification Tolerance}} =$ [ ] %

Notes _____

_____

_____

Analysis performed by: _____ Date: _____

Appendix C

Constants and Formulas
for Control Charts*

$\bar{X}$ and R, $\bar{X}$ and S Charts
Median and Individual Charts

*Adapted from ASTM Publication STP-15D, *Manual on the Presentation of Data Control and Analysis*, 1976, pp. 134–136. Copyright ASTM, 1976. Race Street, Philadelphia, Pennsylvania 19103. Reprinted with permission.

$\overline{X}$ and R, $\overline{X}$ and S Charts

	$\overline{X}$ and R Charts*				$\overline{X}$ and S Charts*			
	Chart for Averages ($\overline{X}$)	Chart for Ranges (R)			Chart for Averages ($\overline{X}$)	Chart for Standard Deviation(s)		
Subgroup Size	Factors for Control Limits	Divisors for Estimate of Standard Deviation	Factors for Control Limits		Factors for Control Limits	Divisors for Estimate of Standard Deviation	Factors for Control Limits	
n	A_2	d_2	D_3	D_4	A_3	c_4	B_3	B_4
2	1.880	1.128	—	3.267	2.659	0.7979	—	3.267
3	1.023	1.693	—	2.574	1.954	0.8862	—	2.568
4	0.729	2.059	—	2.282	1.628	0.9213	—	2.266
5	0.577	2.326	—	2.114	1.427	0.9400	—	2.089
6	0.483	2.534	—	2.004	1.287	0.9515	0.030	1.970
7	0.419	2.704	0.076	1.924	1.182	0.9594	0.118	1.882
8	0.373	2.847	0.136	1.864	1.099	0.9650	0.185	1.815
9	0.337	2.970	0.184	1.816	1.032	0.9693	0.239	1.761
10	0.308	3.078	0.223	1.777	0.975	0.9727	0.284	1.716
11	0.285	3.173	0.256	1.744	0.927	0.9754	0.331	1.679
12	0.266	3.258	0.283	1.717	0.886	0.9776	0.354	1.646
13	0.249	3.336	0.307	1.693	0.850	0.9794	0.382	1.618
14	0.235	3.407	0.328	1.672	0.817	0.9810	0.406	1.594
15	0.223	3.472	0.347	1.653	0.789	0.9823	0.428	1.572
16	0.212	3.532	0.363	1.637	0.763	0.9835	0.448	1.552
17	0.203	3.588	0.378	1.622	0.739	0.9845	0.466	1.534
18	0.194	3.640	0.391	1.608	0.718	0.9854	0.482	1.518
19	0.187	3.689	0.403	1.597	0.698	0.9862	0.497	1.503
20	0.180	3.735	0.415	1.585	0.680	0.9869	0.510	1.490
21	0.173	3.778	0.425	1.575	0.633	0.9876	0.523	1.477
22	0.167	3.819	0.434	1.566	0.647	0.9882	0.534	1.466
23	0.162	3.858	0.443	1.557	0.633	0.9887	0.545	1.455
24	0.157	3.895	0.451	1.548	0.619	0.9892	0.555	1.445
25	0.153	3.931	0.459	1.541	0.606	0.9896	0.565	1.435

$$UCL_{\overline{X}}, LCL_{\overline{X}} = \overline{\overline{X}} \pm A_2\overline{R}$$
$$UCL_R = D_4\overline{R}$$
$$LCL_R = D_3\overline{R}$$
$$\acute{\sigma} = \overline{R}/d_2$$

$$UCL_{\overline{X}}, LCL_{\overline{X}} = \overline{\overline{X}} \pm A_3\overline{s}$$
$$UCL_s = B_4\overline{s}$$
$$LCL_s = B_3\overline{s}$$
$$\acute{\sigma} = \overline{s}/c_4$$

*From ASTM publication STP-15D. *Manual on the Presentation of Data and Control Chart Analysis*, 1976; pp. 134-136.
Copyright ASTM, 1916 Race Street, Philadelphia, Pennsylvania 19103. Reprinted, with permission.

Median and Individual Charts

	Median Charts*,**					Charts for Individuals*				
	Charts for Medians ($\overline{X}$)	Chart for Ranges (R)				Charts for Individuals (X)	Chart for Ranges (R)			
	Factors for Control Limits	Divisors for Estimate of Standard Deviation	Factors for Control Limits			Factors for Control Limits	Divisors for Estimate of Standard Deviation	Factors for Control Limits		
Subgroup Size	$\tilde{A}_2$	d_2	D_3	D_4		E_2	d_2	D_3	D_4	
2	1.880	1.128	—	3.267		2.660	1.128	—	3.267	
3	1.187	1.693	—	2.574		1.772	1.693	—	2.574	
4	0.796	2.059	—	2.282		1.457	2.059	—	2.282	
5	0.691	2.326	—	2.114		1.290	2.326	—	2.114	
6	0.548	2.534	—	2.004		1.184	2.534	—	2.004	
7	0.508	2.704	0.076	1.924		1.109	2.704	0.076	1.924	
8	0.433	2.847	0.136	1.864		1.054	2.847	0.136	1.864	
9	0.412	2.970	0.184	1.816		1.010	2.970	0.184	1.816	
10	0.362	3.078	0.223	1.777		0.975	3.078	0.223	1.777	

$UCL_X, LCL_X = \overline{\overline{X}} \pm \tilde{A}_2\overline{R}$

$UCL_R = D_4\overline{R}$

$LCL_R = D_3\overline{R}$

$\hat{\sigma} = \overline{R}/d_2$

$UCL_X, LCL_X = \overline{\overline{X}} \pm E_2\overline{R}$

$UCL_R = D_4\overline{R}$

$LCL_R = D_3\overline{R}$

$\hat{\sigma} = \overline{R}/d_2$

*From ASTM publication STP-15D. *Manual on the Presentation of Data and Control Chart Analysis.* 1976; pp. 134-136. Copyright ASTM, 1916 Race Street, Philadelphia, Pennsylvania 19103. Reprinted, with permission.

**$\tilde{A}_2$ factors derived from ASTM STP-15D data and efficiency tables contained in W. J. Dixon and F. J. Massey, Jr., *Introduction to Statistical Analysis.* Third Edition, 1969; p. 488; McGraw-Hill Book Company, New York.

Appendix D

Blank Work Forms for the Implementation of Task Analysis Technique (TAT) for Cost of Quality Computations

TASK ANALYSIS FOR COST OF QUALITY

TASK NAME:	
GENERAL DESCRIPTION OF WORK ELEMENTS WITHIN THIS TASK	
ELEMENT CODE	DESCRIPTION
ANALYST NAME:	DEPARTMENT: DATE:
MANAGER'S APPROVAL:	DATE:

TASK ANALYSIS FOR COST OF QUALITY

TASK NAME:
TASK INPUT: _____   _____   _____
INPUT SOURCE: _____   _____
TASK OUTPUT: _____   _____   _____
OUTPUT TO: _____   _____
QUESTION VALUE ADD OF EVERY ELEMENT IN INPUT-OUTPUT KEY AREAS OF   EMPHASIS: _____   _____   _____   _____   _____   _____
ANALYST NAME: DEPARTMENT: DATE:
MANAGER'S APPROVAL: DATE:

TASK ANALYSIS FOR COST OF QUALITY

TASK NAME:

What are your customer requirements for this task?

Are you and your customer in agreement on these requirements?
( ) YES ( ) NO

What problems do poor quality outputs create?

How can/do you measure the quality of the outputs you deliver?

What are the input requirements needed to perform this task?

Are you and your suppliers in agreement on requirements?
( ) YES ( ) NO

What problems do poor quality inputs create?

How can/do you measure the quality of the inputs you receive?

ANALYST NAME:	DEPARTMENT:	DATE:
MANAGER'S APPROVAL:	DATE:	

TASK ANALYSIS FOR COST OF QUALITY

TASK NAME:

How do you monitor and measure the quality of your value add on
this task?

How do you cause poor quality on this task?

What should be done for improvement?

How can improvement be measured?

How many hours/week are spent on this task?   _____hrs/wk

These hours are either a cost of doing business, a cost of
quality, or some combination of the two.
How do you classify them?
                           Business   _____hrs/wk
         Cost of Quality = (COQ) =   _____hrs/wk

COQ hours can be further classified as appraisal, prevention,
and failure. What are they?
                           Appraisal   _____hrs/wk
                           Prevention   _____hrs/wk
                           Failure   _____hrs/wk
                           TOTAL COQ   _____hrs/wk

ANALYST NAME:	DEPARTMENT: DATE:
MANAGER'S APPROVAL:	DATE:

TASK ANALYSIS FOR COST OF QUALITY

IMPROVEMENT PRIORITY #	TASK	APP	PREV	FAIL	NON COQ
ANALYST NAME:		DEPARTMENT:		DATE:	
MANAGER'S APPROVAL:		DATE:			

Appendix E

Productivity and Quality of Working Life Centers in the United States and Abroad*

CENTERS IN THE UNITED STATES

American Center for Quality of Work Life
3301 New Mexico Ave., N.W., Suite 202
Washington, DC 20016
Ted Mills, Director
202/338-2933

Assessment, education, and training in QWL programs for unionized organizations

American Productivity Center
123 N. Post Oak La.
Houston, TX 77024
Dr. C. Jackson Grayson, Chairman 713/681-4020

Public seminars, video-based training programs, information services, and broad range of consulting services

Center for Government and Public Affairs
Auburn University
Montgomery, AL 36117
Dr. Raymond B. Wells, Director
205/279-9110

Technical assistance and consultation to state and local governments in Alabama

---

*Source: Kendrick, J. W., in collaboration with the American Productivity Center: *Improving Company Productivity: Handbook with Case Studies*. The John Hopkins University Press, Baltimore, Maryland 21218, Appendix D. Reprinted with permission.

Center for Manufacturing
Productivity and Technology
Transfer
Jonsson Engineering Center
Rensselaer Polytechnic Institute
Troy, NY 12181
Dr. Leo E. Hanifin, Director
518/270-6000

Solutions to specific
manufacturing problems through
applied engineering and the
transfer of new technology

Center for Productivity and
Quality of Working Life
Utah State University, UMC 35
Logan, UT 84322
Dr. Gary B. Hansen, Director
801/752-4100

Research, education and training,
and consultation on human
resource approaches

Center for Productivity Studies
Kogod College of
Business Administration
The American University
Washington, DC 20016
David S. Bushnell, Director
202/686-2149

Research and consultation on
labor-management and public-
private sector relationships and the
transfer of new technology

Center for Quality of
Working Life
Pennsylvania State University
Capitol Campus
Middletown, PA 17057
Dr. Rupert F. Chisholm, Director
717/787-7746

Education, information, and
assistance on QWL programs for
Pennsylvania organizations

Center for Quality of
Working Life
Institute of Industrial Relations,
UCLA
405 Hilgard Avenue
Los Angeles, CA 90024
Prof. Louis E. Davis, Chairman
213/825-1095

Information, training, and
assistance in organization and job
design approaches

Computer Integrated Design
Manufacturing and Automation
Center
Grissom Hall
Purdue University
West Lafayette, IN 49707
Dean John C. Hancock, Director
317/494-5346

Research into advanced
computer-aided design and
manufacturing systems

Department of Commerce
Productivity Center
U.S. Department of Commerce,
Room 7413
Washington, DC 20230
States L. Clawson, Manager
202/377-3653

Georgia Productivity Center
Georgia Tech Engineering
Experiment Station
Atlanta, GA 30332
Rudolph L. Yobs, Director
404/894-3404

Transfer of technology and safety, environmental, and human resource counseling to Georgia businesses

Harvard Project on Technology, Work, and Character
1710 Connecticut Avenue, N.W.
Washington, DC 20009
Dr. Michael Maccoby, Director
202/462-3003

Research into the relationship between work and human development

Hospital Productivity Center
Texas Hospital Association
P.O. Box 15587
Austin, TX 78761
Dr. Karl L. Shaner
512/453-7204

Dissemination of techniques to improve hospital productivity

Institute for Productivity
592 DeHostos Ave., Baldrich
Hato Ray, PR 00918
Mrs. Milagros Guzman, President
809/764-5145

Research, training, and consulting in human resource approaches to improvement

Laboratory for Manufacturing and Productivity
Massachusetts Institute of Technology, Room 35-136
Cambridge, MA 02139
617/253-2225

Research into innovative manufacturing processes and

systems and development of productivity analysis approaches

Management and Behavioral Science Center
Wharton School
University of Pennsylvania
3733 Spruce St.
Philadelphia, PA 19174
Charles E. Dwyer, Director
215/243-5736

Research and consultation in organizational behavior, job design, and labor-management cooperation

Manufacturing Productivity Center
IIT Center
10 West 35th Street
Chicago, IL 60616
Dr. Keith McKee, Director
312/567-4800

Application of advanced technologies to manufacturing

Maryland Center for Productivity and Quality of Working Life
University of Maryland
College Park, MD 20742
Dr. Thomas C. Tuttle, Director
301/454-6688

Training and education, information services, and assistance in program development for Maryland organizations

National Center for Public Productivity
John Jay College of
Criminal Justice
445 W. 59th Street

New York, NY 10019
Dr. Marc Holzer, Director
212/489-5030

Information clearinghouse, education and training, and technical assistance for the public sector

Northeast Labor-Management Center
30 Church Street
Boston, MA 02178
Dr. Michael J. Brower, Director
617/489-4002

Consultation and training in labor-management and employee involvement programs

Oklahoma Productivity Center
Engineering North
Oklahoma State University
Stillwater, OK 74078
Dr. Earl J. Ferguson, Director
405/624-6055

Training and consultation in both technical and human resource approaches for organizations in Oklahoma and adjoining areas

Oregon Productivity Center
100 Merryfield Hall
Oregon State University
Corvallis, OR 97331
James L. Riggs, Director
503/754-3249

Consulting and technical assistance to small- and medium-sized firms in Oregon

Pennsylvania Technical Assistance Program
J. Orvis Keller Building
University Park, PA 16802
Dr. H. LeRoy Marlow, Director
814/865-0427

Assistance to Pennsylvania organizations in solving specific technical problems

Productivity Center
Chamber of Commerce of the United States
1615 H Street, N.W.
Washington, DC 20062
202/659-3163

Initiatives to influence public policy and organization of conferences

The Productivity Center
University of Miami
P.O. Box 248294
Coral Gables, FL 33124
Dr. David J. Sumanth, Director
305/284-2344

Seminars and publications

Productivity Council of the Southwest
5151 State University Drive, STF 124
Los Angeles, CA 90032
John R. Frost, Director
213/224-2975

Information, education, and consultation to organizations in southwestern region

Productivity Evaluation Center
302 Whitmore Hall
Virginia Tech
Blacksburg, VA 24061
Dr. P. M. Ghare, Director
703/961-6656

Education and program
development in industrial
engineering and participative
approaches

Productivity Institute
College of Business
Administration
Arizona State University
Tempe, AZ 85281
Dr. Eileen Burton, Director
602/965-7626

Research, information, training,
and problem identification services

**Productivity Research and
Extension Program**
North Carolina State University
P.O. Box 5511
Raleigh, NC 27607
Dr. William A. Smith, Director
919/733-2370

Information and operations
improvement projects with
emphasis on manufacturing

**Puerto Rico Economic
Development Administration**
G.P.O. Box 2350
San Juan, PR 00936

Quality of Working Life Program
Ohio State University
1375 Perry Street
Columbus, OH 43201
Dr. Don Ronchi, Director
614/422-3390

Research and consultation in
labor-management cooperation

**State Government Productivity
Research Center**
Council of State Governments
P.O. Box 11910
Lexington, KY 40578
James E. Jarrett, Director
606/252-2291

Information on state government
projects and productivity
approaches nationwide

**Texas Center for Productivity and
Quality of Work Life**
Texas Tech University
P.O. Box 4320
Lubbock, TX 74909
Barry A. Macy, Director
806/742-1538

Information and consultation on
work innovation and
organizational change

Work in America Institute
700 White Plains Road
Scarsdale, NY 10583
Jerome M. Rosow, President
914/472-9600

Education and training, technical
assistance, and information on
work practices

INTERNATIONAL AND FOREIGN COUNTRY CENTERS

International and Regional

Arab League Industrial
Development Organization
El-Phairr Square
Cairo, Egypt

Asian Productivity Organization
4-14 Akasaka 8-Chome
Minato-Ku, Tokyo, 107 Japan
Hiroshi Yokota, Secretary-
General
Contact: Mr. George C. Shen,
Director of Administration and
Public Relations

European Association of National
Productivity Centres
Rue de la Concorde, 60
1050 Brussels, Belgium
Contact: Mr. A. C. Hubert,
Secretary-General

Inter-American Productivity
Association
Casilla 13120
Santiago, Chile
Contact: Mr. Kovacevitz

International Labor Organization
Geneva, Switzerland

United Nations Industrial
Development Organization
Lerchenfelderstrasse #1, A-1070
Vienna, Austria
Contact: H. Abdul-Rahman,
Secretary-General

Country Programs by Region

Europe

Denmarks Erhversfond
Codamhus-Gl. Kongevej 60
DK-1850 Copenhagen V.,
Denmark
Contact: P. Assam, Director

British Council of Productivity
Associations
8 Southhampton Row
London WC1B 4AQ, England
Mr. D. F. Bailey, Chief Executive

British Institute of Management
Management House Parker Street
London WC 2B 5PT, England
Roy Close, Director General
Correspondence contact: William
Bree, Deputy Director General

Finland Foundation of
Productivity Research, Federation
of Finnish Industries
Etelaranta 10
00130 Helsinki 13, Finland

Centre National d'Information
pour le Progress Economique
(CNIPE)
Tour Europe, Cedex 07
92080 Paris, La Defense, France

Greek Productivity Center
28, Copodistriou Street
Athens, Greece
Contact: D. Talellis

Hungarian Institute for Industry
and Industrial Education Society
V, Szechenyi rakpart 3
Budapest, Hungary

Industrial Development Institute
of Iceland
Skipholt 37
Reykjavik, Iceland
Contact: S. Bjornsson

European Foundation for the
Improvement of Living and
Working Conditions
Loughlinstown House
Shankhill, co. Dublin, Ireland
Contact: Mrs. W. O. Conghaile

Irish Productivity Centre
IPC House, 35–39
Shelbourne Road
Dublin 4, Ireland
Contact: J. Ryan

Instituto Nazionale per
l'Incremento della Produttivita
(INIP)
Piazza Indipendenza 11/8
00185 Rome, Italy
Contact: S. Cimmino

Office Luxembourgeois pour
l'Accroissement de la Productivité
(OLAP)
18, rue Auguste Limiere,
Luxembourg
Contact: J. Faltz

Commissie Opvoering
Produktiviteit Van de Sociaal-
Economische Radd
Bezuidenhoutseweg 60
Den Haag, The Netherlands
Contact: C. A. M. Mul

Norsk Produktivitetsinstitutt
(NPI)
Postboks 8401
Hammersborg
Oslo 1, Norway
Contact: S. Dalen, Director

Instytut Przemyslu U
Drobnego i Rzemiosla
aleje Jerozolimskire 87
02-0001 Warsaw, Poland

Conselho Nacional de
Produtividade
Avenido Don Carlos I 126-3
Lisbon, Portugal

Subdireccion General de
Technologia y Productividad
Industrial
Ayala, 3
Madrid 1, Spain

National Productivity Center
(MPM) 46 Mithatpasa Caddesi
Yenisehir
Ankara, Turkey
Contact: A. Ulubay,
Secretary-General

Turkey Training and
Productivity Centre
P.O. Box 554 Nicosia
via Mersin 10, Turkey
Contact: H. M. Ateskin, Director

Jugoslpvenski Zavoda Za
Produktivnost Rada
1, Uzun Mirkova
Belgrade, Yugoslavia
Contact: V. Odovic, Director

Rationalisierungskuratorium der
Deutschen Wirtschaft (RKW)
Dusseldorf Strasse
6236 Eschborn, West Germany
Dr. V. Ruhle, professor

North America (Not Including United States)

Canada Department of Industry,
Trade and Commerce
Productivity Branch
235 Queen Street
Ottawa, Ontario
Canada K1A 0H5
Contact: Dr. Imre Bernolak

Department of Regional Economic
Expansion
43rd Floor, 800 Place Victoria
Montreal, Quebec
Canada H42 1E8
Contact: Roger Fournier

Institut National de Productivité
51 rue d'Auteuil
Quebec, Canada G1R 4C2
Correspondence contact:
Jacinthe Belisle

Centro Nacional de Productividad
de Mexico, A.C.
Anillo Periferico Sur 2143
Mexico 20, D.F.
Correspondence contact:
Dr. Gustavo Polit, Programa
Editorial e Informacion
Internacional

Monterrey Productivity Center
Edificio de las Instituciones
Campo 250, Pte 4° Plso
Monterrey, Nuevo Leon, Mexico
Contact: Hugo Mier Arrieta

Central and South America

National Productivity Center
of Brazil
Ministry of Industry, Commerce
and Labor
Brazilia, Brazil

Servicio de Cooperacion Tecnica
Huerfanos 1117-1147, 9° piso
Casilla 276, Correo Central
Santiago, Chile

Centro De Desaroolo Industrial
del Ecuador (CENDES)
Avenida Orellana 1297
P.O. Box 2321
Quito, Ecuador

Instituto Tecnico de Capicitacion
y Productividad (Intecap)
12 Calle 4–17
Zone 1
Guatemala City, Guatemala

Centro de Desarrollo y
Productividad Industrial de
Panama
Ministerio de Industrie
Apartado 7639
Panama City, Panama 5

Centro Nacional de Productividad
Presidencia de la Republica
Pablo Bermudez 214, Piso 10
Lima, Peru
Contact: George Succar, Director

Fundacion Instituto Venezalano de
Productividad
Avenida Libertador
Edificio Nuevo Centro-Chacao
Caracas, Venezuela

Asia

East Asia

Hong Kong Productivity Centre
20th & 21st Floors, Sincere Bldg.
173 Des Voeux Road
P.O. Box 16132
Central Hong Kong, Hong Kong
Dr. J. C. Wright,
Executive Director
Correspondence contact:
Ms. Cangidi Chan, Public
Relations Officer

Japan Productivity Center
No. 1-1. Shibuya 3-Chome
Shibuya-ku
Tokyo 150, Japan
K. Goshi, Chairman

Productivity and Development
Center
Development Academy of the
Philippines (IDP-DAP)
P.O. Box 5160
Makati Rizal, The Philippines
Mr. Arturo L. Tolentino,
Managing Director

National Productivity Board
6th Floor, Cuppage Centre
55 Cuppage Rd.,
Singapore 0922
Dr. Goh Keng Leng,
Executive Director
Contact: Mrs. Lau Wai Ying,
Administrative Manager

Korea Productivity Center
10, 2-ka, Pil-Dong
Chung-Ku
Seoul, South Korea
Eun Bok Rhee, President
Contact: Mr. Kwan Won Rhim,
Managing Director

China Productivity Center
11th Floor, 201-26 Tun Hua
North Rd.
Taipei, Taiwan 100
Contact: Mr. Wang Sze-Cheh,
General Manager

Thailand Management
Development and Productivity
Centre
Department of Industrial
Promotion
Ministry of Industry
Government of Thailand
Rama 6 Road
Bangkok 4, Thailand
Mr. Thamnu Vasinonta, Director

South Asia

National Productivity Council
Productivity House
5–6 Institutional Area
Lodi Road
New Delhi, India 11003
Mr. K. T. Chandy, Chairman
Correspondence contact:
S. L. Mehta, International
Projects & Services Officer

National Productivity Centre
Ministry of Manpower,
Transmigration and Cooperatives
Jalan Letjen Haryono M.T.
(Kantor Ditjen Transmigrasi)
P.O. Box 358-kby
Jakarta Selatan, Indonesia
Mr. Rusli Syarif, Director

Industrial Management Institute
Bahlavie Road
Jame Jam Avenue
Tehran, Iran
Jamshid Gharajedaghi,
Managing Director

Nalaysia National Productivity
Centre
P.O. Box 64
Yalan Sultan
Petaling Yaya, Malaysia

Industrial Services Centre
Post Box-1318
Kathmandu, Nepal
Mr. Ajit N. S. Thapa, Chairman
Correspondence contact:
Keshab P. Sharma, Chief
Industrial Extension Services
Division

Pakistan Industrial Technical
Assistance Centre
Maulana Jalal-Ud-Din Roomi
Road
Post Office, La hore 16, Pakistan
Brig. M. A. Faruqui,
General Manager

Economic Affairs Division
Ministry of Planning & Economic
Affairs
P.O. Bos 1689
Colombo, Sri Lanka
G. P. H. Leelananda de Silva,
Senior Assistant Secretary &
Director

Sri Lanka National Institute
of Management
7, Kollupitya Station Road
Calombo-3, Sri Lanka

Australasia

Australia Department of
Productivity
Ministry of Industry, Anzac Park
West Building
Constitution Avenue
Parkes A.C.T. 2600, Australia
Contact: Vernon White

Nepean Productivity Center
P.O. Box 10
Kingswood 2750, Australia
Contact: Dr. M. R. Ramsay,
director

Productivity Promotion Council
of Australia
Head Office 339 Swanston Street
GPO Box 475D
Melbourne, Victoria 3000, Australia
Contact: Mr. D. L. Casey,
Executive Director

Department of Trade and Industry
Productivity and Technology
Division
Private Bag
Wellington, New Zealand

Middle East

Israel Institute of Productivity
4 Szold Street
Tel-Aviv, Israel
Dr. Israel Meidan,
Executive Director
Correspondence contact:
Ms. G. Simon, Information
Officer

Syria Management Development
and Productivity Centre
Abdlghani Tollo Bldg.
Mohajrin
Damascus, Syria

Africa

Ghana Management Development
and Productivity Institute
P.O. Box 297
Accra, Ghana

Nigeria Federal Ministry of
Productivity and Employment
1 Koyi, Lagos, Nigeria

National Productivity Institute
P.O. Box 3971
Pretoria 0001
Republic of South Africa
Dr. Jan Visser, Executive Director

Sudan Management Development
and Productivity Centre
P.O. Box 2308
Khartoum, Sudan

Tunisia Institut National
de Productivité
Rue bel Hassen Ben Chaabane
el Omrane
Tunis, Tunisia

Sources for Additional Information on Productivity and Innovation Centers in the United States and Around the World

Directory of U. S. Productivity and Innovation Centers (Publication 6233), published by the Chamber of Commerce of the United States, 1615 H Street N.W., Washington, D.C. 20062

International Directory of Productivity Centers, published by Case and Text Administration, Room 29, School of Business and Administration, The University of Western Ontario, London, Canada N6A3K7

Appendix F

Glossary of Terms

Accuracy: deviation of the measured or observed value from the true value (*see also* Precision)

Advanced statistical methods: more sophisticated and less widely applicable techniques of statistical process analysis and control than included in basic statistical methods; this can include more advanced control chart techniques, regression analysis, design of experiments, and advanced problem-solving techniques

Allocation criteria: used for allocating overhead expenses to the various input and output components

Arithmetic mean: often referred to as the average of the population

Attributes data: qualitative data that can be counted for recording and analysis; examples include characteristics, such as the presence of a required label, the installation of all required fasteners, the absence of errors on an expense report; other examples are characteristics that are inherently measurable (may be treated as variables data), but for which the results are recorded in a simple yes/no fashion, such as acceptability of a shaft diameter when measured on a go/no-go gage, or the presence of any engineering changes on a drawing; attributes data are usually gathered in the form of nonconforming units or of nonconformities; they are analyzed by p, np, c, and u control charts (*see also* Variables data)

Average: sum of values divided by the number (sample size) of values; designated by a bar over the symbol for the values being averaged: $\bar{X}$ (X bar) is the average of the X values within a subgroup; $\bar{\bar{X}}$ (X double bar) is the average of subgroup averages; $\tilde{\bar{X}}$ (X tilde bar) is the average of subgroup media; $\bar{p}$ (p bar) is the average of p values from all the subgroups (*see also* Mean)

Awareness: personal understanding of the interrelationship of quality and productivity, directing attention to the requirement for management commitment and statistical thinking to achieve never-ending improvement

Basic statistical methods: apply the theory of variation through use of basic problem-solving techniques and statistical process control; include control chart construction and interpretation (for both variables and attributes data) and capability analysis

Binomial distribution: discrete probability distribution for attributes data that applies to conforming and nonconforming units and underlies the p and np charts

Block: part of the experimental material that is likely to be more homogeneous than the whole

Capability: (can be determined only after the process is in statistical control); when the process average plus and minus and 3σ spread of the distribution of individuals ($\bar{\bar{X}} + 3\sigma$) is contained within the specification tolerance (variables data), or when at least 99.73% of individuals are within specification (attributes data), a process is said to be capable; efforts to improve capability must continue, however, consistent with the operational philosophy of never-ending improvement in quality and productivity

Capital: in the context of the present study, the investment in natural resources, reproducible capital (structures, machinery, equipment, and inventories), and financial assets, excluding investments in government debt and in securities of other enterprises

Capital compensation: income accruing to owners of property in the form of interest, rent, royalties, and profits

Capital consumption: using up stored services and the resulting decline in value of reproducible durable capital as a result of aging, deterioration, and obsolescence; not to be confused with capital input, which represents the use of extant capital goods

Capital, fixed: sum of the value of land, structures, machinery, and equipment

Capital, working: sum of the value of cash, accounts and notes receivable, and inventories

Cause-and effect diagram: simple tool for individual or group problem-solving that uses a graphic description of the various process elements to analyze potential sources of process variation; also called fishbone diagram (after its appearance) or Ishikawa diagram (after its developer)

Central line: the line on a control chart that represents the average or median value of the items being plotted

Characteristic: a distinguishing feature of a process or its output on which variables or attributes data can be collected

Common cause: a source of variation that affects all the individual values of the process output being studied; in control chart analysis it appears as part of the random process variation

Comparative experiments: an experiment whose objective is to compare the treatments rather than to determine absolute values

Comprehensive productivity planning cycle (CPPC): formalized process for planning productivity in an organization; has four major components: productivity planning appraisal process (PPAP), strategic productivity planning process (SPPP), tactical productivity planning process (TPPP), and operational productivity planning process (OPPP)

Comprehensive quality planning: process of assessing all factors that may affect product or service quality throughout the specified life cycle

Computer-integrated manufacturing (CIM): fully integrated CAD/CAM system that provides computer assistance from marketing to product shipment

Consecutive: units of output produced in succession; a basis for selecting subgroup samples

Continuous improvement in quality and productivity: the operational philosophy that makes best use of the talents within the company to produce products of increasing quality in an increasingly efficient way that protects the return on investment

Control chart: a graphic representation of a characteristic of a process, showing plotted values of some statistic gathered from that characteristic, a central line, and one or two control limits; minimizes the net economic loss from type I and type II errors; has two basic uses: as a judgment to determine if a process has been operating in statistical control, and as an operation to aid in maintaining statistical control

Control limit: a line (or lines) on a control chart used as a basis for judging the significance of the variation from subgroup to subgroup; variation beyond a control limit is evidence that special causes are affecting the process; control limits are calculated from process data and are not to be confused with engineering specifications

Deflation (price): dividing an economic time series expressed in value terms by an index of prices of the underlying physical units (combined by appropriate quantity weights) in order to convert the series to "real" terms or constant prices

Deflator: generally an index of price used to bring the current dollar output and/or input(s) to base-period terms; a deflator can be less than 1.0, although, because of inflation, it is usually greater than 1.0

Delphi technique: qualitative forecasting technique that utilizes a panel of experts and a series of questionnaires to develop a forecast

Detection: a past-oriented strategy that attempts to identify unacceptable output after it has been produced and then separate it from the good output (*see also* Prevention)

Distribution: a way of describing the output of a common-cause system of variation in which individual values are not predictable but in which the outcomes as a group form a pattern that can be described in terms of its location, spread, and shape; location is commonly expressed by the mean or average, or by the median; spread is expressed in terms of the standard deviation or the range of a sample; shape involves many characteristics, such as symmetry and peakedness, but these are often summarized by using the name of a common distribution, such as the normal, binomial, or Poisson

Effect of a factor: change in response produced by a change in the level of the factor (applicable only for factors at two levels each)

Experiment: a planned set of operations that leads to a corresponding set of observations

Experimental unit: one item to which a single treatment is applied in one replication of the basic experiment

Factor-independent variable: a feature of the experimental conditions that may be varied from one observation to another; may be qualitative or quantitative, fixed or random

Forecasting: a tool used in productivity and quality planning process; various forecasting techniques, such as exponential smoothing and arithmetic average, are useful in the planning process

Group technology: the organizing and planning of production parts into batches that have some similarity of geometry and/or processing sequence

Harmonization: the integration of the interests of the stockholders (owners), the board of directors, management, and employees in a consistent manner both within and outside the physical boundaries of an organization

Human factors engineering (ergonomics): the multidisciplinary activity of designing human/equipment interface to match the human capacities with the physical work environment

Index number: device for measuring proportionate changes or differences in simple or complex quantities relative to their "base" magnitude; index numbers of a time series, the most common type, represent magnitudes in given periods as percentages of their value in a base period

Index-number problem: differences in movement of a quantity (price) index resulting from the use of different weight bases if there is a systematic relationship between relative changes in quantities sold and prices

Individual: a single unit or a single measurement of a characteristic

Interaction: if the effect of one factor is different at different levels of another factor, the two factors are said to interact or to have interaction

Levels of a factor: the various values of a factor considered in the experiment are called levels

Location: general concept for the typical values or central tendency of a distribution

Main effect: the average effect of a factor is called the main effect of the factor

Mean: the average of values in a group of measurements

Median: the middle value in a group of measurements, when arranged from lowest to highest; if the number of values is even, by convention the average of the middle two values is used as the median; subgroup medians form the basis for a simple control chart for process location; medians are designated by a tilde (˜) over the symbol for the individual values: $\tilde{X}$ is the median of a subgroup

Morale management: ongoing process of improving factors that affect employee productivity and quality

Motivation: motivational strength to satisfy a need

Noncomparative experiment: an experiment whose objective is the determine of the properties or characteristics of a population

Nonconforming units: units that do not conform to a specification or other inspection standard; sometimes called discrepant or defective units; p and np control charts are used to analyze systems producing nonconforming units

Nonconformities: specific occurrences of a condition that does not conform to specifications or other inspection standards, sometimes called discrepancies or defects; an individual nonconforming unit can have the potential for more than one nonconformity (a door may have several dents and dings; a functional check of a carburetor may reveal any of a number of potential discrepancies); c and u control charts are used to analyze systems producing nonconformities

Normal distribution: a continuous, symmetrical, bell-shaped frequency distribution for variables data that underlies the control charts for variables; when measurements have a normal distribution, about 68.26% of all individuals lie within ± 1 standard deviation unit of the mean, about 95.44% lie within ± 2 standard deviation units of the mean, and about 99.73% lie within ± 3 standard deviation units of the mean. These percentages are the basis for control limits and control chart analysis (since subgroup averages are normally distributed even if the output as a whole is not), and for many capability decisions (since the output of many industrial processes follows the normal distribution)

Operational definition: means of clearly communicating quality expectations and performance; it consists of (1) a criterion to be applied to an object or to a group, (2) a test of the object or of the group, (3) a decision: yes or no—the object or the group did or did not meet the criterion

Operational productivity planning process (OPPP): concerns itself with detailed implementation of the plans

Outcome (response)-dependent variable: result of a trial with a given treatment is called a response

Pareto diagram: a simple tool for problem-solving that involves ranking all potential problem areas or sources of variation according to their contribution to cost or to total variation; typically, a few causes account for most of the cost (or variation), so problem-solving efforts are best prioritized to concentrate on the "vital few" causes, temporarily ignoring the "trivial many"

Poisson distribution: discrete probability distribution for attributes data applies to nonconformities and underlies the c and u control charts

Precision: a measurement's precision is related to its repeatability in terms of the deviation of a group of observations from a mean value; the terms "accuracy" and "precision" are often used interchangeably, but they may be distinguished as accuracy being the measure of the approach to a true value, and precision a measure of consistency or repeatability

Prevention: a future-oriented strategy that improves quality and productivity by directing analysis and action toward correcting the process itself; consistent with a philosophy of never-ending improvement

Problem-solving: process of moving from symptoms to causes (special or common) to actions that improve performance; among the techniques that can be used are Pareto charts, cause-and-effect diagrams, and statistical process control techniques

Process: combination of people, equipment, materials methods, and environment that produce output—a given product or service; can involve any aspect of a business; a key tool for managing processes is statistical process control

Process average: location of the distribution of measured values of a particular process characteristic, usually designated as an overall average $\bar{X}$

Process control: a state whereby statistical inferences techniques are used to monitor and control a specified process in order to achieve improved quality and gains in productivity

Process control system: a feedback mechanism that provides information about process characteristics and variables, process performance, action on the process, inputs, transformation process, and action on the output

Process spread: extent to which the distribution of individual values of the process characteristic vary; often shown as the process average plus or minus some number of standard deviations (such as $\bar{\bar{X}} + 3\sigma$)

Production: process of transforming resources (inputs) into products (outputs) that satisfy human wants; sometimes used as a synonym for output, which is the result of the production process

Production and service improvement techniques (PASIT): ongoing process that involves organized use of common sense to find easier and better ways of performing work and streamlining the production and service processes to ensure that goods and services are offered at minimum overall cost

Production worker: defined by the Bureau of Labor Statistics to include workers (up through the working foreman level) engaged in fabricating, processing, assembling, inspecting, receiving, storing, handling, packing and warehousing, shipping (but not delivering), maintaining, and repairing, as well as janitorial work, security services, product development, auxiliary production for the plant's own use, record-keeping, and other services closely associated with these production operations

Productivity (capital): the ratio of total output to capital input

Productivity (computer operating expenses): the ratio of total output to computer operating expenses input

Productivity (data processing expenses): the ratio of total output to data processing expenses input

Productivity (energy): the ratio of total output to energy input

Productivity (labor): the ratio of total output to labor input

Productivity management triangle (PMT): presents the components of productivity management, measurement, evaluation and control, planning and analysis, and improvement and monitoring

Productivity (materials): the ratio of total output to materials input

Productivity (other administrative expenses): the ratio of total output to other administrative expenses input

Productivity (partial): the ratio of total output to one class of input

Productivity planning appraisal process (PPAP): PPAP provides the basis for understanding the threats, opportunities, strengths, and weaknesses in the environment in which the organization operates

Productivity and quality challenges: several issues that deserve significant attention while attempting to improve productivity and quality

Productivity and quality connection: integrated relationship between productivity and quality

Productivity and quality coordinator: company official charged with the responsibility of coordinating productivity and quality improvement program

Productivity and quality evaluation and control: evaluating the productivity and quality indicators of an organization, task, plant, etc.

Productivity and quality indicator: a factor, issue, or indicator that impacts on productivity and quality

Productivity and quality management: integrated process involving both management and employees with the ultimate goal of managing the design, development, production, transfer, and use of the various types of products or services in both the work environment and marketplace; requires total involvement of everyone in the planning, measurement, evaluation, control, and improvement of productivity and quality at the source of production or service center

Productivity and quality management checklist: a comprehensive self-assessment instrument for identifying areas of weaknesses that deserve productivity and quality improvement in an organization

Productivity and quality management hierarchy (PQMH): presents the levels and the basis to identify the productivity and quality issues from a global perspective

Productivity and quality management program: a program design to specifically improve productivity and quality within an organization

Productivity and quality management triangle (PQMT): a formalized process of productivity and quality measurement, planning and analysis, evaluation and control, improvement and monitoring

Productivity and quality measurement: process of measuring the productivity and quality at different levels such as at the task level, organizational level, or national level

Productivity and quality planning: planning both productivity and quality variables within an organization, task, plant, etc.

Productivity and quality problems: several factors affecting productivity level and growth rate and the quality of goods and services

Productivity (robotics operating expenses): the ratio of total output to robotics operating expenses input

Productivity (total): the ratio of total output to all input factors

Productivity (total-factor): the ratio of total output to the sum of associated labor and capital factor inputs

Profitability: product of a effective and efficient productivity and quality management program

Profitability analysis: explanation of changes in profits in terms of changes in total productivity and in "priced recovery," defined as the ratio of prices received for outputs to prices paid for inputs

Quality: defined as "fitness for use" by Juran; "conformance to specification" by Crosby; "fitness for use in terms of the ability to process and produce with less rework, less scrap, minimal down time and high productivity" by Edosomwan

Quality characteristics: elements of fitness to use that typify the variety of uses of a given product

Quality circle (QC): small voluntary group of coworkers from a plant or office who meet periodically to formulate and solve problems and thus raise productivity; in Japan, where this approach has been highly developed, the circle members are given special training in various problem-solving techniques

Quality cost (appraisal): costs resulting from the extra effort expanded to assure conformance to quality standards and performance specifications

Quality cost (external): costs generated by defective products shipped to customers

Quality cost (internal): costs associated with defective products and discovered prior to product delivery to the customer

Quality cost (prevention): costs resulting from the prevention of nonconformance items

Quality costs: various categories of costs that associated with identifying, avoiding, producing, maintaining, and repairing products do not meet specifications

Quality error removal (QER) technique: provides a framework, principles, and guidelines to a group of employees who voluntarily work together to select and solve key problem(s), affecting an organization's work unit or task; organizational goals are broken down into small tasks at the operational level, and continuous effort is applied to improve productivity and quality within each work unit

Randomness: condition in which individual values are not predictable, although they may come from a definable distribution

Range: difference between the highest and lowest values in a subgroup; the expected range increases both with sample size and with the standard deviation

Repeatability: describes measurement variation obtained when one person measures the same dimension or characteristic several times with the same gage or test equipment (sometimes referred to as "equipment variation")

Reproducibility: a term popularized in the automotive industry as representing the variation in measurement averages when more than one person measures the same dimension or characteristic using the same measuring instrument

Response: the numerical result of a trial based on a given treatment combination

Run: a consecutive number of points consistently increasing or decreasing, or above or below the central line; can be evidence of the existence of special causes of variation

Run chart: simple graphic representation of characteristics of a process, showing plotted values of some statistics gathered from the process (often individual values) and a central line (often the median of the values), which can be analyzed for runs

Sample: in process control applications, a synonym for subgroup; this use is totally different from the purpose of providing an estimate of a larger group of people or items

Shape: general concept for the overall pattern formed by a distribution of values

Sigma (σ): Greek letter used to designate a standard deviation

Special cause: source of variation that is intermittent, unpredictable, and unstable; sometimes called an assignable cause; signaled by a point beyond the control limits or a run or other nonrandom pattern of points within the control limits

Specification: engineering requirement for judging the acceptability of a particular characteristic; a specification is never to be confused with a control limit

Spread: general concept for the extent to which values in a distribution differ from one another; dispersion

Stability (for control charts): the absence of special causes of variation; the property of being in statistical control

Stability (for gage studies): variation in the measurement averages when the measuring instrument values are recorded over a specified time interval

Stable process: a process that is in statistical control

Standard deviation: a measure of the spread of the process output or the spread of a sampling statistic from the process (of subgroup averages); denoted by the Greek letter σ (sigma)

Statistic: a value calculated from or based upon sample data (a subgroup average or range), used to make inferences about the process that produced the output from which the sample came

Statistical control: condition describing a process from which all special causes of variation have been eliminated and only common causes remain; evidenced on a control chart by the absence of points beyond the control limits and by the absence of nonrandom patterns or trends within the control limits

Statistical process control: use of statistical techniques, such as control charts, to analyze a process or its outputs to take appropriate actions to achieve and maintain a state of statistical control and to improve the process capability

Strategic productivity planning process (SPPP): enables long-range goals and objectives to be formulated

Subgroup: one or more events or measurements used to analyze the performance of a process; rational subgroups are usually chosen so that the variation represented within each subgroup is as small as feasible for the process (representing the variation from common causes), and so that any changes in the process performance (special causes) will appear as differences between subgroups; rational subgroups are typically made up of consecutive pieces, although random samples are sometimes used

Tactical productivity planning process (TPPP): concerns itself with the means of implementing the goals and objectives specified in SPPP

Task-oriented total productivity measurement (TOTPM) model: model for measuring the total productivity of a task and the organization as a whole

Technology-oriented total productivity measurement (TOTPM) model: model for measuring the total productivity of a technology, project, product, product group, and the organization as a whole

Total horizontal job enlargement and enrichment (THJEAE): a process involving both employees and management in expanding job responsibilities to include a greater variety of activities

Total vertical job enlargement and enrichment (TVJEAE): a process involving both employees and management in planning, organizing, performing, and improving the job content

Treatment combination: set of levels of all factors included in a trial in an experiment is called a treatment or treatment combination

Type I error: rejecting an assumption that is true; taking action appropriate for a special cause when in fact the process has not changed; overcontrol

Type II error: failing to reject an assumption that is false; not taking appropriate action when in fact the process is affected by special causes; undercontrol.

Utility index: defined by Stewart as a "surrogate measure" to produce a single number

Variables data: quantitative data, where measurements are used for analysis; examples include the diameter of a bearing journal in millimeters, the closing effort of a door in kilograms, the concentration of electrolyte as a percentage, or the torque of a fastener in Newton-meters; $\bar{X}$ and R, $\bar{X}$ and s, median and individual control charts are used for variables data (*see also* Attributes Data)

Variation: inevitable differences among individual outputs of a process; the sources of variation can be grouped into two major classes: common causes and special causes

Weight: indicator of relative importance, such as prices, by which physical units of outputs or of inputs are combined to provide aggregate measures

Weight base: period from which relative weights are drawn; may or may not be the same as the "comparison base" from which values are set equal to 100.0 for index number construction

Worker participation: approach to overcoming resistance to change through employee involvement in the planning and implementation of the change

Working condition improvement: a technique that involves a detailed audit of working conditions at each operation, designing improved working conditions, and installing and maintaining improvements in the working conditions

Work measurement: a means of determining an equitable relationship between the quantity of work performed and the number of labor hours required for completing the quantity of work

Work simplification: systematic investigation and analysis of contemplated and present work systems and methods for the purpose of developing easier, quicker, less fatiguing, and more economical ways of providing high-quality goods and services

Zero defects: a quality philosophy that strives for virtually error-free production of goods and services

Appendix G

Bibliography

Abell, D. F., 1980. *Defining the Business.* Prentice Hall, Englewood Cliffs, New Jersey.

Ackoff, R. L., 1981. *Creating the Corporate Future.* John Wiley and Sons, New York.

Adam, E. E., Jr., J. C. Hershauer, and W. A. Ruch, 1978. *Measuring the Quality Dimension of Service Productivity.* National Science Foundation, U.S. Department of Commerce, Washington, D.C.

Adam, E. E., Jr., J. C. Hershauer, and W. A. Ruch, 1981. *Productivity and Quality: Measurement as a Basis for Improvement.* Prentice-Hall, Englewood Cliffs, New Jersey.

Adam, N. R., and A. Dogramaci, Eds., 1981. *Productivity Analysis at Organizational Level.* Kluwer Boston, Hingham, Massachusetts.

Aggarwal, S. C., 1979. A study of productivity measures for improving benefit-cost ratios of operating organizations. Proceedings 5th International Conference Production Research, Amsterdam, The Netherlands, August 12–16, pp. 64–70.

American National Standards Institutes, 1975. *Control Chart Method of Controlling Quality During Production* (ASQC Standard B3-1958/ANSI Z1.3-1958, revised.

American National Standards Institute, 1975. *Guide for Quality Control and Control Chart Method of Analyzing Data* (ASQC Standards B1-1958 and B2-1958/ANSIZ1.1-1958 and Z1.2-1958, revised.

American Productivity Center, 1978. Productivity and the industrial engineer. Region VIII, AIIE Conference, Chicago, October 27, (seminar notes).

343

American Society for Quality Control, 1976. *QC Circles: Applications, Tools, and Theory,* 161 West Wisconsin Avenue 53203.

American Society for Testing and Materials, 1976. *Manual on Presentation of Data and Control Chart Analysis* (STP-15D).

American Society for Testing Materials, 1951. *ASTM Manual on Quality Control of Materials,* Philadelphia.

Andrews, K. R., 1971. *The Concept of Corporate Strategy.* Dow Jones/Irwin, Homewood, Illinois.

Ansoff, H. I., 1965. *Corporate Strategy: An Analytic Approach to Business Policy for Growth and Expansion.* McGraw-Hill, New York.

Arrow, K. J., H. B. Chenery, B. S. Minhas, and R. M. Solow, 1961. Capital-labor substitution and economic efficiency. Rev. Econ. Stat., Vol. XVIII, pp. 225-250.

Barnes, R. M., 1968. *Motion and Time Study Design and Measurement of Work,* 6th Ed. John Wiley and Sons, New York.

Bemesderfer, J. L., 1979. Approving a process for production. *Journal of Quality Technology,* January, pp. 1-12.

Berger, W., 1978. Micro computers and software quality control. ASQC Annual Technical Conference Transactions, p. 328.

Bernolak, I., 1976. Enhancement of productivity through interfirm comparisons, a Canadian experience. In *Improving Productivity Through Industry and Company Measurement,* Series 2. National Center for Productivity and Quality of Working Life, U.S. Govt. Printing Office, Washington, D.C., pp. 59-65.

Blanchard, K. and S. Johnson, 1982. *The One Minute Manager.* William Morrow, New York.

Bowen, W., 1979. Better prospects for our ailing productivity. *Fortune,* December 3, pp. 68-76.

Box, G. E. P., W. G. Hunter, and S. J. Hunter, 1978. *Statistics for experimenters.* John Wiley and Sons, New York.

Bradley, J. W., and D. H. Korn, *Acquisition and Corporate Development: A Contemporary Perspective for the Manager.* Arthur D. Little, Lexington, Massachusetts.

Bright, J. R., 1968. *Technological Forecasting for Industry and Government.* Prentice-Hall, Englewood Cliffs, New Jersey.

Cetron, M. J., 1969. *Technological Forecasting; A practical Approach.* Technology Forecasting Institute, New York.

Cetron, M. J., and A. L. Weiser, 1968. Technological change, technological forecasting and planning R&D—a View from the R&D manager's desk. *The George Washington Law Review Technology Assessment and the Law,* Vol. 36, No. 5, July, pp. 1090, 1091.

Charbonneau, H. C., and G. L. Webster, 1978. *Industrial Quality Control.* Prentice-Hall, Englewood Cliffs, New Jersey.

Cole, R. E., 1981. The Japanese lesson in quality. *Technology Review,* July, p. 29.

Cotton, F., 1976. In productivity, planning is everything. *Industrial Engineering,* November.

Craig, C. E., and C. R. Harris, 1972. Productivity concepts and measurement—a management viewpoint. Unpublished Master's Thesis. MIT, Cambridge, Massachusetts.

Craig, C. E., and C. R. Harris, 1973. Total productivity measurement at the firm level. *Sloan Management Review,* Vol. 14, No. 3, pp. 13–39.

Crandall, N. F., and L. M. Wooton, 1978. Development strategies of organizational productivity, *California Management Review,* Vol. 21, No. 2, pp. 37–46.

Crosby, P. B., 1979. *Quality is Free.* Signet, New York.

Dalkey, N. C., 1969. The Delphi method: An experimental study of group opinion. The RAND Corporation, RM-5888 PR, June, Santa Monica, California.

Danforth, D. D., 1984. Quality means doing the job right the first time. *The Wall Street Journal,* March 21, p. 33.

Davies, O. L., 1954. *Design and Analysis of Industrial Experiments.* Oliver and Boyd, London.

Davis, H. S., 1955. *Productivity Accounting.* University of Pennsylvania Press, Philadelphia.

Deming, E. W., 1982. *Quality, Productivity, and Competitive Position.* MIT Press, Cambridge, Massachusetts.

Dewar, D. L., 1980. *The Quality Circle Handbook.* Quality Circle Institute, Red Bluff, California.

Dewitt, F., 1970. Technique for measuring management productivity. *Management Review,* Vol. 59, pp. 2–11.

Dewitt, F., 1976. Productivity and the industrial engineer. *Industrial Engineering,* Vol. 8, No.1, pp. 20-27.

Dhrymes, P. J., 1963. Comparison of productivity behavior in manufacturing and service industries. Rev. Econ. Stat., Vol. 45, No. 1, pp. 64-69.

Diamond, W. J. *Practical Experiment Designs for Engineers and Scientists.* Lifetime Learning Publications, Division of Wadsworth, Inc., Belmont, California.

Domar, E. D., et al., 1964. Economic growth and productivity in the United States, Canada, United Kingdom, Germany and Japan in the post-war period. Rev. Econ. Stat., Vol. 46, pp. 33-40.

Doyle, M., and D. Straus, 1978. *How to Make Meetings Work.* Playboy Paperbacks, New York.

Duncan, A. J., 1974. *Quality Control and Industrial Statistics,* Richard D. Irwin, Homewood, Illinois.

Edosomwan, J. A., 1980. Implementation of the total productivity model in a manufacturing company. Master's Thesis, Department of Industrial Engineering, University of Miami, July.

Edosomwan, J. A., 1983a. Production and service improvement technique (PASIT). Unpublished manual. IBM Data Systems Division, New York.

Edosomwan, J. A., 1983b. Quality error removal technique. Unpublished Manual, IBM Data Systems Division, New York.

Edosomwan, J. A., 1985a. A methodology for assessing the impact of computer technology on productivity, production quality, job satisfaction and psychological stress in a specific assembly task. Doctoral Dissertation, Department of Engineering Administration, The George Washington University, Washington, D.C. 20052, January. Grant SS-36-83-21, Social Science Research Council (U.S. Department of Labor) and IBM 2J2/2K5/-722271/83/85.

Edosomwan, J. A., 1985b. A task-oriented total productivity measurement model for electronic printed circuit board assembly. International Electronic Assembly Conference Proceeding, October 7–9, Santa Clara, California.

Edosomwan, J. A., 1985c. Computer-aided manufacturing impact on productivity, production quality, job satisfaction, and psychological stress in an assembly task. Proceedings of the Annual International Industrial Engineering Conference, December.

Edosomwan, J. A., 1985d. Quality at the source of production. IBM technical working paper.

Edosomwan, J. A., 1986a. A conceptual framework for productivity planning. *Industrial Engineering,* January.

Edosomwan, J. A., 1986b. The impact of computer-aided manufacturing on total productivity. Proceedings for the 8th Annual Conference on Computers and Industrial Engineering, Orlando, Florida, March.

Edosomwan, J. A., 1986c. Managing technology in the workplace: A challenge for industrial engineers. *Industrial Engineering,* February, pp. 14-18.

Edosomwan, J.A., 1986d. A methodology for assessing the impact of robotics on total productivity in an assembly task. Proceedings of Annual International Industrial Engineering Conference, May, Dallas, Texas.

Edosomwan, J. A., 1986e. Productivity management in computer-aided manufacturing environment. Proceedings of the First International Conference on Engineering Management, September, Washington, D.C.

Edosomwan, J. A., 1986f. Productivity and quality management—a challenge in the year 2000. Proceedings of Annual International Industrial Engineering Conference, December 7–10, Boston, Massachusetts.

Edosomwan, J. A., 1986g. Robotics-aided task impact model for the new frontier in manufacturing. Proceedings of Second World Conference on robotics Research. Robotic Research transactions published by Society of Manufacturing Engineers (SME).

Edosomwan, J. A., 1986h. Statistical process control in group technology production environment. SYNERGY '86 Proceedings, June 16–18, Universal City, California: Sponsored by Society of Manufacturing Engineers, Computer and Automated Systems Association, and the American Production and Inventory Control Society.

Edosomwan, J. A., 1986i. Technology impact on the quality of working life—a challenge for engineering managers in the year 2000. Proceedings of the First International Conference on Engineering Management, September, Washington, D.C.

Edosomwan, J. A., 1987. The meaning and management of productivity and quality. *Industrial Engineering*, January.

Edosomwan, J. A., and D. J. Sumanth, 1985. *A Practical Guide for Productivity Measurement in Organizations*. Working Manual.

Ewing, D. W., ed., 1964. *Long-Range Planning for Management*. Harper and Row, New York.

Fabricant, S., 1969. *A Primer on Productivity*. Random House, New York.

Fabricant, S., 1962. Which Productivity? Perspective on a current question. *Monthly Labor Review*, Vol. 86, No. 6, pp. 609-613.

Farago, F. T., 1982. *Handbook of Dimensional Measurement*. Industrial Press, New York.

Feigenbaum, A. V., 1979. American manufacturers strive for quality—Japanese style. *Business Week,* March 12, p. 5.

Feigenbaum, A. V., 1961. *Total Quality Control*. McGraw-Hill, New York.

Fein, M., 1974. *Rational Approaches to Raising Productivity*. Monograph Series No. 5, American Institute of Industrial Engineering, Norcross, Georgia.

Fein, M., 1976. *Designing and Operating an IMPROSHARE Plan*. Hillsdale, New Jersey, July 2.

Fetter, R. B., 1967. *The Quality Control System*. Richard D. Irwin, Homewood, Illinois.

Fiske, T. S., 1981. Program utilizes learning and the length of run to determine tightness of line balance. *Industrial Engineering*, August, pp. 58-62.

Forrester, J. W., 1961. *Industrial Dynamics*. MIT Press, Cambridge, Massachusetts.

Geare, A. J., 1976. Productivity from Scanlon-type plans. Academic Management Review, Vol. 1, No. 3, pp. 99-107, July.

Gold, B., 1976. Tracing gaps between expectations and results of technological innovation: The case of iron and steel. *Journal of Industrial Economics*, September,

Goodwin, H. F., 1968. Improvements must be managed. *Journal of Industrial Engineering,* No. 11, pp. 538-543.

Grant, E. L., and R. S. Leavenworth, 1980. *Statistical Quality Control*, 5th Ed. McGraw-Hill, New York.

Greene, J. H., 1970. *Production and Inventory Control Handbook*. McGraw-Hill, New York.

Greene, J. H., 1974. *Production and Inventory Control: Systems and Decisions,* revised Ed. Richard D. Irwin, Homewood, Illinois.

Groff, G. K., 1971. Worker productivity: An integrated view. *Business Horizons,* Vol. 14, pp. 78-86, April.

Groff, G. K., and T. B. Clark, 1981. Commentary on production operations management: Agenda for the '80s. Decis. Sci., Vol. 12, pp. 573-581.

Groocock, J. M., 1974. *The Cost of Quality.* Pitman, New York.

Gryna, F. M., Jr., 1977. Quality costs: User vs. manufacturer. *Quality Progress,* Vol. 10, No. 6, June, pp. 10–15.

Hahn, G. J., and S. S. Shapiro, 1968. *Statistical Models in Engineering.* John Wiley and Sons, New York.

Ham, I., 1976. Introduction to group technology. Technical Report MMR76-03. Society of Manufacturing Engineers, Dearborn, Michigan.

Harrigan, K. R. *Strategies for Declining Businesses.* Lexington Books, D. C. Heath and Company, Lexington, Massachusetts.

Harris, F. K., 1962. *Electrical Measurements.* John Wiley and Sons, New York.

Haveman, R. H., and G. Christainsen, 1982. *Jobs and the Environment.* Work in America Institute, Scarsdale, New York.

Helmer, O., 1969. Analysis of the future: The Delphi method. In *Technological Forecasting for Industry and Government,* J. R. Bright, ed. Prentice-Hall, Englewood Cliffs, New Jersey.

Henderson, B. D., 1982a. *Henderson on Corporate Strategy.* Mentor Books, New York.

Henderson, B. D., 1982b. Just in time (JIT). *Perspectives,* Vol. 244. The Boston Consulting Group, Boston.

Hershauer, J. C., and W. A. Ruch, 1978. A worker productivity model and its use at Lincoln Electric. *Interfaces,* Vol. 8, No. 3, pp. 80-90.

Herzberg, F., 1968. One more time: How do you motivate employees? *Harvard Business Review,* Vol. 46, No. 1, pp. 53-62, January/February.

Herzberg, F., B. Mansner, and D. B. Snyderman, 1959. *The Motivation to Work.* John Wiley and Sons, New York.

Hicks, C. R., 1956. Fundamentals of analysis of variance, Parts I, II and III. *Industrial Quality Control,* August, September, October.

Hines, W. W., 1976. Guidelines for implementing productivity measurement. *Industrial Engineering,* Vol. 8, No. 6.

Holusha, J., 1984. Quality Woes Bedevil Detroit, Gain Made, But Japan Is Still Seen as Leader. *The New York Times,* April 30, p. D-1.

IBM, 1973. *Communications Oriented Production Information and Control System (COPICS),* Vols. I-VIII, White Plains, New York.

IBM Data Systems Division, *Teaming up for Quality: Quality Excellence Teams Education Guide.* Poughkeepsie, New York 12602.

IBM Quality Institute, 1985. *Process Control, Capability and Improvement.* The Quality Institute, Southbury, Connecticut, May.

Ishikawa, K., 1976. *Guide to Quality Control,* Revised Ed. Asian Productivity Organization, Tokyo.

Jehring, J. J., 1967. A contrast between two approaches of total systems incentives. *California Management Review,* Vol. 10, No. 1, pp. 7-14.

Jones, H., 1974. *Preparing Company Plans: A Workbook for Effective Corporate Planning.* John Wiley and Sons, New York.

Jucius, J. M., 1963. *Personnel Management.* Richard D. Irwin, Homewood, Illinois.

Juran, J. M., 1964. *Managerial Breakthrough.* McGraw-Hill, New York.

Juran, J. M., 1973. The Taylor system and quality control, a series of eight papers. *Quality Progress* (American Society for Quality Control), May to December.

Juran, J. M., 1978. Life behind the quality dikes. European Organization for Quality Control, 22nd Annual Conference, Dresden.

Juran, J. M., 1979. *Quality Control Handbook.* McGraw-Hill, New York.

Juran, J. M., and F. N. Gryna, Jr., 1970. *Quality Planning and Analysis.* McGraw-Hill, New York.

Juran, J. M., and F. M. Gryna, Jr., 1980. *Quality Planning and Analysis.* McGraw-Hill, New York.

Juran, J. N., 1981. Product quality—A presentation for the west. *Management Review*, American Management Association, June/July, p. 16.

Juran, J. M., F. M. Gryna, Jr., and R. S. Bingham, Jr., 1979. *Quality Control Handbook,* 3rd Ed. McGraw-Hill, New York.

Juran, J. M., F. M. Gryna, Jr., 1980. *Quality Planning and Analysis,* McGraw-Hill, New York.

Kapur, R., and D. H. Liles, 1982. Job design for persons with physical disabilities. AIIE Proceedings Annual Spring Conf., May, pp. 169-178.

Kendrick, J. W., in collaboration with the American Productivity Center, 1984. *Improving Company Productivity. Handbook with Case Studies.* The John Hopkins University Press, Baltimore.

Kendrick, J. W., and D. Creamer, 1965. *Measuring Company Productivity: Handbook with Case Studies* (Studies in Business Economics, No. 89). National Industrial Conference Board, New York.

Khalil, T. M., 1976. The role of ergonomics in increasing productivity. AIIE Proceedings Annual Spring Conference, pp. 57-64.

Konz, S., 1979. Quality Circles: Japan success story. *Industrial Engineering,* Vol. 11, No. 10, pp. 24-28.

Kroemer, K. H. E., and D. L. Price, 1982. Ergonomics in the office: Comfortable work stations allow maximum productivity. *Industrial Engineering,* pp. 24-32, July.

Latham, G. P., 1981. *Increasing Productivity Through Performance Appraisal.* Addison-Wesley, Reading, Massachusetts.

Lawler, E. E., III, 1971. *Pay and Organizational Effectiveness: A Psychological Review.* McGraw-Hill, New York.

Lee, E., and M. B. Packer, 1981. The uses of productivity information in the private business sector. Working paper, Laboratory for Manufacturing and Productivity, MIT, Cambridge, Massachusetts, October.

Lee, M. D., and J. R. Hackman, 1982. *Redesigning Work: A Strategy for Change.* Work in America Institute, Scarsdale, New York.

Leek, J. W. and F. H. Riley, 1978. Product quality improvement through visibility. ASQC Annual Technical Conference Transactions, 229-236.

Lenz, R. C., Jr., 1966. Technological forecasting. Paper presented at the U.S. Air Force Symposium on Long-Range Forecasting and Planning, Colorado Springs, Colorado, August, pp. 155-157.

Lenz, R. C., Jr., 1968. Forecast of exploding technologies by trend extrapolation. In *Technical Forecasting for Industry and Government,* J. R. Bright, Ed. Prentice-Hall, Englewood Cliffs, New Jersey, pp. 65-69.

Lorange, P., 1980. *Corporate Planning: An Executive Viewpoint.* Prentice-Hall, Englewood Cliffs, New Jersey.

Loup, R. J., M. K. Murphy, and C. L. Russell, 1978. Quality control in a health information system. ASQC Annual Technical Conference Transactions, pp. 451-457.

Mali, P., 1972. *Managing by Objectives.* John Wiley and Sons, New York.

Mali, P., 1978. *Improving Total Productivity: MBO Strategies for Business Government, and Not-for-Profit Organizations.* John Wiley and Sons, New York.

Malzahn, D., 1982. Job modification and placement strategies for persons with physical disabilities using the available motions inventory. AIIE Proceedings Annual Spring Conference, May, pp. 179-187.

Mao, J. C. T., 1965. Measuring productivity of public urban renewal expenditures. *Michigan Business Review,* vol. 17, pp. 30–34.

Maslow, A., 1943. A theory of human motivation. *Psychological Review,* July, pp. 388-389.

Mayer, R. R., 1982. *Production and Operations Management,* 4th Ed. McGraw-Hill, New York.

McClelland, D. C., 1961. *The Achieving Society.* Van Nostrand, Princeton.

McCormick, E. J., 1976. *Human Factors in Engineering and design.* McGraw-Hill, New York.

McGee, J. P., 1981. Job design technology in duPont to improve productivity and job satisfaction. AIIE Proceedings Annual Spring Conference, pp. 461-464, May.

McGregor, D., 1960. *The Human Side of Enterprise.* McGraw-Hill, New York.

Melman, S., 1956. *Dynamic Factors in Industrial Productivity.* John Wiley and Sons, New York.

Midas, M. T., 1981. The productivity-quality connection. *Design News,* December 7, pp. 56.

MIL-C-45662A, 1962. *Military Specification: Calibration System Requirements.* U.S. Government Printing Office, Washington, D.C.

Mills, C. A., 1976. In-plant quality audit. *Quality Progress,* October, pp. 23-25.

Montgomery, D. C., 1976. *Design and Analysis of Experiments.* John Wiley and Sons, New York.

Morris, W. T., 1977. *Productivity Measurment Systems for Administrative Computing and Information Services,* Vol. 1. National Science Foundation, Washington, D. C.

Morrison, D. L., and K. E. McKee, 1978. Technology for improved productivity *Manufacturing Productivity Frontier,* Vol. 2, No. 6, June, pp. 1-6.

Mundel, M. E., 1970. *Motion and time Study Principles and Practice.* Prentice Hall, Englewood Cliffs, New Jersey.

Mundel, M. E., 1976. Measures of productivity. *Industrial Engineering,* Vol. 8, No. 5, pp. 24-26.

Muramatsu, R., 1981. Example of increasing productivity and product quality through satisfying the workers' desires and developing the workers' motivation. AIIE Proceedings Annual Spring Conference, May, pp. 652-660.

Natrella, M. G., 1966. *Experimental Statistics.* NBS Handbook 91. U.S. Government Printing Office, Washington, D. C., October.

OEEC, 1950. *Terminology of Productivity.* Par 2,2 rue Andre-Pascal, Paris-16.

Ohmae, K., 1982. *The Mind of the Strategist: The Art of Japanese Business.* McGraw-Hill, New York.

Omachonu, V. K., 1980. Productivity improvement: Conceptual framework, model and implementation methodology for manufacturing companies. Master's Thesis, Department of Industrial Engineering, University of Miami, Coral Gables, Florida, July.

Ott, E. R., 1975. *Process Quality Control.* McGraw-Hill, New York.

Ouchi, W. G., 1981. *Theory Z: How American Business Can Meet the Japanese Challenge.* Avon Books, New York.

Pearce and Robinson, 1982. *Formulation and Implementation of Competitive Strategy.* Richard D. Irwin, Homewood, Illinois.

Plackett, R. L., and J. P. Burmen, 1946. The design of multifractorial experiments. *Biometrika,* Vol. 33, pp. 305-325.

Polglase, D. G., 1981. Productivity growth through inter-firm cooperation. AIIE Proceedings Annual Spring Conference, May, pp. 772-779.

Porter, M., 1980. *Competitive Strategy: Techniques for Analyzing Industries and Competitors.* Free Press, New York.

Quality Cost-Cost Effectiveness Technical Committee, American Society for Quality Control, 1971. *Quality Costs—What and How,* 2nd Ed.

Quality Magazine, 1981. Quality, a management gambit. June.

Quinn, J. B., 1967. Technological forecasting. *Harvard Business Review,* Vol. 45, No. 2.

Quinn, J. B., 1980. *Strategies for Change: Logical Incrementalism.* Richard D. Irwin, Homewood, Illinois.

Rahn, R. W., et al., 1981. *Productivity, People, and Public Policy.* U.S. Chamber of Commerce, Washington, D.C.

Robinson, M., 1982. *Quality Circle, A Practical Guide.* Gower Publishing, Aldershot, England.

Rodriquez, R., and O. Adaniya, 1985. Group technology cell allocation. 1985 Annual International Industrial Engineering Conference Proceedings.

Roscoe, J. T., 1969. *Fundamental Research Statistics for Behavioral Sciences.* Holt, Rinehart and Winston, New York.

Rothschild, W. E., 1979. *Strategic Alternatives: Selection, Development and Implementation.* Amacom, New York.

Ruch, W. A., and J. C. Hershauer, 1974. *Factors Affecting Worker Productivity.* Arizona State University, Tempe.

Schonberger, R. J., 1982. *Japanese Manufacturing Techniques: Nine Hidden Lessons in Simplicity.* Free Press, New York.

Seder, L. A., and D. Cowan, 1956. *The SPAN Plan Method for Process Capability Analysis.* ASQC General publication 3, September.

Shilliff, K. A., and M. Bodis, 1975. How to pick the right vendor. *Quality Progress,* January, pp. 12-14.

Shue, L., 1981. Productivity improvement through quality control. AIIE Proceedings Annual Spring Conference, May, pp. 793-799.

Siegel, I. H., 1976. Measurement of company productivity in improving productivity through industry and company measurement (National Center for Productivity and Quality of Working Life, Series 2). U.S. Government Printing Office, Washington, D. C.

Siegel, I. H., 1980. *Company Productivity: Measurement for Improvement.* The W. E. Upjohn Institute for Employment Research, Kalamazoo, Michigan, April.

Sink, D. S., 1982. Motivating employees in the 80's for increased productivity and quality. Proceedings IIE Annual Spring Conference. May, pp. 324-334.

Sink, S., 1985. Strategic planning: A crucial step toward a successful productivity management program. *Industrial Engineering,* January, pp. 52-60.

Smith, A., 1776. *The Wealth of Nations.*

Smith, L. A., and J. L. Smith, 1982. How can an IE justify a human factors activities program to management? *Industrial Engineering,* pp. 39-43, February.

Special Report, 1982. Quality: The U.S. drives to catch up. *Business Week,* November 1.

Statistical Engineering Laboratory, 1957. *Fractional Factorial Experiment Designs for Factors at Two Levels.* NBS Applied Mathematics Series 48, April.

Stewart, W. T., 1978. A yardstick for measuring productivity. *Industrial Engineering,* Vol. 10, No. 2, pp. 34-37.

Sumanth, D. J., 1979. Productivity measurement and evaluation models for manufacturing companies. Doctoral Dissertation, Illinois Institute of Technology, Chicago, August. (University Microfilms, Ann Arbor, Michigan, No. 80-03, 665).

Sumanth, D. J., 1984. *Productivity Engineering and Management.* McGraw-Hill, New York.

Sutermeister, R. A., 1976. *People and Productivity,* 3rd. Ed. McGraw-Hill, New York.

Sweetland, J., 1982. *Occupational Stress and Productivity.* Work in America Institute, Scarsdale, New York.

Taylor, B. W., III, and Davis, R. K., 1977. Corporate productivity—getting it all together. *Industrial Engineering,* Vol. 9, No. 3, pp. 32-36.

Taylor, F. W., 1911. *Scientific Management.* Harper and Brothers, New York.

Toeppner, T. G., 1970. Implementing a corporate quality assurance activity in a multi-product, divisionalized corporation. ASQC Technical Confererence Transaction, pp. 1-12.

Tucker, S. A., 1961. *Successful Management Control by Ratio Analysis,* McGraw-Hill, New York.

Turner, J. A., 1980. Computers in bank clerical functions: Implication for productivity and the quality of working life. Doctoral Dissertation, Columbia University, New York.

U.S. Chamber of Commerce, 1981. *The World Economy: Recent Trends and Outlook.* Economic Policy Division, U.S. Government Printing Office, Washington, D. C., August.

U.S. Department of Commerce, 1977. *Computer Software Management: A primer for project Management and Quality Control.* Washington, D.C.

U.S. Department of Labor, Bureau of Labor Statistics, 1980. *Monthly Labor Review,* January, pp. 40–43.

Vroom, V. H., 1964. *Work and Motivation.* John Wiley and Sons, New York.

Walker, R. C., 1950. The problem of the repetitive job. *Harvard Business Review,* Vol. 28, No. 3, pp. 54–58.

Warnecke, H.-J., H.-J. Bullinger, and J. H. Kolle, 1981. German manufacturing industry approaches to increasing flexibility and productivity. AIIE Proceedings Annual Spring Conference, May, pp. 643-651.

Western Electric Co., 1956. *Statistical Quality Control Handbook.* Available from I.D.C. Commercial Sales, Western Electric Company, PO Box 26205, Indianapolis, Indiana 46226.

Wilson, M. F., 1967. The quality your customer sees. *Journal of the Electronics Division ASQC,* July, pp. 3-16.

Index